WITHDRAWN
UTSA Libraries

SESSION INITIATION PROTOCOL (SIP)

Controlling Convergent Networks

About the Author

Travis Russell has been in telecommunications more than 25 years, working in both data and voice telephony networks. He has held numerous positions from field engineer to marketing and technical sales. Today Travis is a Senior Manager with Tekelec, where he has spent the last 18 years of his career working in SS7, SIP, and now IMS. He has filed numerous patents and appeared as speaker at numerous industry events worldwide. Travis is also the author of several other McGraw-Hill books, including *Signaling System #7,* and *The IP Multimedia Subsystem (IMS): Session Control and Other Network Operations.*

SESSION INITIATION PROTOCOL (SIP)

Controlling Convergent Networks

Travis Russell

New York   Chicago   San Francisco   Lisbon   London   Madrid
Mexico City   Milan   New Delhi   San Juan   Seoul
Singapore   Sydney   Toronto

The McGraw·Hill Companies

Library of Congress Cataloging-in-Publication Data

Russell, Travis.
  Session Initiation Protocol (SIP) : controlling convergent networks /
Travis Russell.
     p. cm.
  Includes bibliographical references and index.
  ISBN 978-0-07-148852-5 (alk. paper)
  1. Computer network protocols. 2. Internet telephony. 3. Wireless
communication systems. I. Title.
  TK5105.55.S36 2008
  621.382'12—dc22

                                        2008018678

McGraw-Hill books are available at special quantity discounts to use as premiums and sales promotions, or for use in corporate training programs. To contact a special sales representative, please visit the Contact Us page at www.mhprofessional.com.

Session Initiation Protocol (SIP): Controlling Convergent Networks

Copyright © 2008 by The McGraw-Hill Companies. All rights reserved. Printed in the United States of America. Except as permitted under the Copyright Act of 1976, no part of this publication may be reproduced or distributed in any form or by any means, or stored in a database or retrieval system, without the prior written permission of publisher.

1234567890 DOC DOC 0198

ISBN   978-0-07-148852-5
MHID   0-07-148852-9

Sponsoring Editor
  Jane K. Brownlow

Editorial Supervisor
  Jody McKenzie

Project Manager
  Harleen Chopra,
  International Typesetting
  and Composition

Acquisitions Coordinator
  Jennifer Housh

Copy Editor
  Bob Campbell

Proofreader
  Manish Tiwari

Indexer
  Broccoli Information
  Management

Production Supervisor
  George Anderson

Composition
  International Typesetting
  and Composition

Illustration
  International Typesetting
  and Composition

Art Director, Cover
  Jeff Weeks

Cover Designer
  12E Design

Information has been obtained by McGraw-Hill from sources believed to be reliable. However, because of the possibility of human or mechanical error by our sources, McGraw-Hill, or others, McGraw-Hill does not guarantee the accuracy, adequacy, or completeness of any information and is not responsible for any errors or omissions or the results obtained from the use of such information.

Library
University of Texas
at San Antonio

Behind every man is a woman, friend, and confidant who helps us wake up every morning and begin our day with a purpose in life. I have been blessed with such a woman who has remained my friend and partner in life for the last 26 years.

She has stood by my side through good and bad, and has always believed in everything I have pursued. She is the mother to my three beautiful children who have grown to be astonishing young adults, each in his or her own unique way, through her example and influence.

Thanks, Deby, for being such a wonderful wife, and for raising such wonderful kids.

Contents at a Glance

Chapter 1   Architecture of a SIP Network 1

Chapter 2   Structure of the SIP Protocol 21

Chapter 3   SIP Status Codes 55

Chapter 4   Registration Procedures in a SIP Network 73

Chapter 5   Establishing a Session in SIP 91

Chapter 6   Extending SIP to Support New Functions 115

Chapter 7   Security in a SIP Network 133

Appendix A   SIP-Related RFCs 159

Appendix B   Methods and Parameters 163

Appendix C   Methods and Parameters from
                a Proxy Perspective 207

Bibliography 247

Index 251

Contents

Acknowledgments xiii
Introduction xv

Chapter 1   Architecture of a SIP Network **1**

The Traditional Voice Network 2
    Wireline Network Architecture 2
    Wireless Network Architecture 5
Network Elements in a Voice over IP Network 7
    Media Gateway (MG) 9
    Media Gateway Control Function (MGCF) 10
    Signaling Gateway 10
    Application Servers (ASs) 11
    The Domain Name System (DNS) 11
    Electronic Numbering (ENUM) 12
SIP-Specific Entities 13
    User Agents (UAs) 14
    Proxy Servers 16
    Redirect Servers 18
    Registrars 19
    Location Servers 19

Chapter 2   Structure of the SIP Protocol **21**

SIP Messages and Formats 21
    Concept of a Dialog 23
    Requests 25
    Responses 26
    Header Fields , ... 28
SIP Identities 42
    Private User Identity 44
    Public User Identity 45

Session Description Protocol (SDP) 46
    Session Descriptions 48
    Time Descriptions 50
    Media Descriptions 51
    Attributes 52

Chapter 3   SIP Status Codes **55**

1*xx* Provisional Codes 57
2*xx* Successful Status Codes 59
3*xx* Redirection Status Codes 60
4*xx* Client Failure Status Codes 61
5*xx* Server Failure Status Codes 69
6*xx* Global Failure Status Codes 70

Chapter 4   Registration Procedures in a SIP Network **73**

Basic Registration 74
Event Notification 77
    Message Waiting Indication (MWI) 77
Interworking with the PSTN 79

Chapter 5   Establishing a Session in SIP **91**

Accessing the Network 92
Initiating a Dialog 93
Client Request 97
Server Response 99
    Emergency Session Establishment 101
SIP Routing 102
    Loose Routing 103
    Strict Routing 108
SIP Session Modification 111
SIP Session Termination 112

Chapter 6   Extending SIP to Support New Functions **115**

The Concept of SIP Extensions 116
    How Extensions are Documented 117
    How Extensions are Treated 117
Some Examples of Extensions 118
    P-Access-Network-Info 119
    P-Answer-State 120
    P-Asserted-Identity 121
    P-Associated-URI 121
    P-Called-Party-ID 122
    P-Charging-Function-Addresses 122
    P-Charging-Vector 123

P-Early Media 124
P-Media-Authorization 125
P-Preferred-Identity 125
P-Profile-Key 126
P-User-Database 126
P-Visited-Network-ID Header 127
Packet Cable Extensions 128
P-DCS-Trace-Party-ID 128
P-DCS-OSPS 128
P-DCS-Billing-Info 129
P-DCS-LAES 130
P-DCS-Redirect 131

Chapter 7   Security in a SIP Network 133

Types of Network Attacks 134
Registration Hijacking 135
Session Hijacking 136
Impersonating a Server 137
Tampering with Message Bodies 138
Tearing Down Sessions 138
Denial of Service and Amplification 139
Bots and DDoS Attacks 140
Security Measures 141
Password and Access Controls 145
Encryption 146
Authentication and Authorization 149
Strict Routing 150
Security Solutions 151
Intrusion Detection 152
Intrusion Protection 156

Appendix A   SIP-Related RFCs 159

IETF SIP Requests for Comments (RFCs) 159

Appendix B   Methods and Parameters 163

ACK Method 164
BYE Method 165
CANCEL Method 170
INVITE Method 172
MESSAGE Method 178
NOTIFY Method 182
OPTIONS Method 186
REGISTER Method 191
SUBSCRIBE Method 196
UPDATE Method 201

Appendix C   Methods and Parameters from a Proxy Perspective 207

ACK Method 208
BYE Method 209
CANCEL Method 213
INVITE Method 215
MESSAGE Method 220
NOTIFY Method 224
OPTIONS Method 229
REGISTER Method 233
SUBSCRIBE Method 238
UPDATE Method 243

Bibliography 247

Index 251

Acknowledgments

A project such as this doesn't happen without the influence of individuals, the encouragement from friends and family, and the support of partners in the project. This book is no exception.

Many thanks to all my colleagues at Tekelec who have encouraged me over all these years to continue writing books. There are many individuals at Tekelec who have been a constant source of encouragement, too many to name here. Thanks to all of you for continuing to support me and encourage me to continue writing.

This makes number seven, and it has been particularly rewarding because it has allowed me a chance to research something new, and to use my experience from packet networks. There will be many more pages in editions to come, as technology continues to evolve and mature. On that note, thanks to Jane Brownlow, my editor, for working with me through this project, and Steve Chapman for his ideas and concepts on how this book should go.

Of course, no author could make it through without the support of his family. My daughters have grown up understanding my commitment to writing and have always been very supportive even if it meant I was buried in my office in specs and drafts. Thanks, Nichole and Laura, for always having something positive to say and for always understanding when Dad had to finish another chapter.

There were many nights when I would be sitting in the camper writing instead of standing around the campfire. My son, Travis Jr., often asked what I was doing sitting inside when there was so much going on outside, but deadlines are seldom forgiving, and sacrifices have to be made. Thanks, son, for understanding your Dad's deadlines. Stoke the fire, because the manuscript is done and Dad's coming out!

Introduction

Telephone networks have come a long ways in the last 100 years. Or have they? If one looks at how a telephone network works, not much has changed. Connections are still made through "dedicated" circuits connected end to end for the duration of a call. While switchboard operators have been replaced by digital switching systems, the concept of connecting those circuits has not evolved much.

The telephone itself is still the same boring instrument, consisting of a speaker, a microphone, and a dial. About the most innovation we have seen there is the addition of a touch-tone dial and a display for caller names.

The networks themselves are still hierarchical switching networks, although some of the layers have been collapsed over the years. Calls are still routed through dedicated routes based on the digits dialed, and a tandem switch is still used to interconnect to other networks. This is a far cry from packet networks.

Packet networks have been around for many years, dedicated to transporting data from network to network. I can remember working on packet networks and building circuits during my tenure at the Bell System. I was fortunate that I was one of the few who worked on both voice and data networks because of my background in data and my expertise in data and voice networks.

Back then packet networks were used for dedicated computing purposes, providing the backbone for major computer networks. Public access was not provided, and the concept of connecting to the Internet did not yet exist (there was no public Internet at the time). In fact, there weren't many personal computers then. Apple had emerged with the Apple II (I traded in my Apple I for the Apple II).

I remember when I first got a "portable" computer. It was one of the first "laptops," if you will, although you would need a pretty big lap for this beast. It was an Osborne Computer, designed to be transportable (it had a really big handle) so that you could compute on the move. What that really meant was you could haul it from work to home and set it up on the kitchen table.

Of course, in those days there were no hot spots, Starbucks didn't yet exist, and the only wireless telecommunications was radiotelephone. Yet even then, the concept of marrying the computer with the telephone was there. My Osborne had an RJ-11 jack on the back where you could plug in a telephone line for the modem that allowed you to connect the computer at a lightning-fast 300 baud.

The software on the computer included a program that allowed you to look up telephone numbers and dial them from the computer. A handset connected to the back of the computer then allowed you to use the computer as your phone (perhaps the first address book?). But the voice itself was still analog and did not go through the data network. In fact, the data connection was made through the telephone network.

Then as years moved on, digital switching became more prevalent and we began seeing systems designed to support both voice and data. Of course, voice was always the primary function of the switches, with the added capability of packet switching. I remember the first true digital packet switch I worked with was a thing called the Lexar, built by Lexar Corporation (later to become United Technologies, then Telex, than Memorex Telex, and then I lost track).

The voice was actually packetized and sent to digital phones, which meant that data could also be packetized and sent to and from the phones. Computers connected via a serial connector (remember those?) on the back of the phone at a dizzying speed of 19.2 Kbps. We were getting closer to seeing computers and data combined, but still, there lacked a network that could bridge all of these systems together.

There were, of course, mainframe computers and dedicated networks running SNA. I did work on some of these for a while, but they were so inflexible and the SNA protocol was very limited compared to today's protocols. The network elements were pretty basic, operating at the transport layer to route data packets from mainframe to terminal. These networks were really designed to provide dumb terminals access to the mainframes, rather than to share data between users.

Then came the Internet, and everything changed. Suddenly we all had access to a giant public network that allowed us to connect and exchange all types of data with anyone who had a connection. New Internet service providers (ISPs) began popping up offering electronic mail and message boards, and later news services and newsgroup access.

These ISPs continue to grow and offer new services as the Internet matured. When the World Wide Web (WWW) was introduced, communication was changed forever. Suddenly we could shop online, establish our own presence on the Internet, and socialize with people from all over the world.

It was only natural that the next big service would be Voice over IP (VoIP). It is very fulfilling for me to see these networks finally maturing to where they have become mainstream after so many decades of experimentation and failed attempts. Today we can finally enjoy the many features and capabilities that packet networks enable with telephone service combined with our e-mail and data service.

Yet for many years it was like the Wild Wild West, with many different implementations and many different configurations to support VoIP networking. There were the legacy telephony guys trying to make the packet network emulate a switched network, and the data guys trying to make the packet network support voice.

In the end, we meet in the middle. New protocols have been developed to help make VoIP more robust and reliable while supporting quality as good as or superior to legacy networks. We see networks beginning to mature and standardize on their technologies so that they can interoperate with other networks, and so vendor equipment will interoperate end to end.

And now, the Session Initiation Protocol (SIP) emerges from the dust as the winner for session control, at least as far as the legacy service providers are concerned. The fact that the 3rd Generation Partnership Project (3GPP) has defined SIP as the standard for call session control for their IP Multimedia Subsystem (IMS) architecture is not minor. There is significant work underway today to add to the SIP protocol to enable it to provide services like we have never seen before.

This book attempts to capture most of the SIP protocol as it has been defined today. This includes the baseline RFC 3261, and many of the extensions that have followed to support new services in the network using the SIP protocol. As SIP continues to mature and grow, I will capture those changes in subsequent editions of this book, so this has become an ongoing work in progress.

I have attempted to simplify SIP as much as possible, and focus on the baseline functions rather than go into great detail about the protocol and its procedures. There is quite a bit of detail here, without going into the nitty-gritty developer stuff. Hopefully this will help everyone to better understand how SIP works, and how to make SIP work in your own network.

Architecture of a SIP Network

Over the last few decades, there have been very few dynamic changes to the telecommunications networks of the world. Telephones were once connected by an operator who used a hardwire cable to connect one telephone wire to another, bridging the connection between subscribers. As years went by, these "switchboards" became larger in size, with many more connections to be made as the subscriber base grew.

The telephone switch was invented as a method of connecting these telephone wires without human intervention, but the method for making those connections really was no different than the switchboard concept. The difference was that the connection was being made through electromechanical relays instead of an operator.

The basic concept of making a connection did not change. The bridge was being made electromechanically; the concept of bridging two wires together to form an end-to-end circuit had not changed. As years went by, this principle remained the same, albeit the networks became more complex and sophisticated through multiple layers of switches and facilities.

The proper term for this type of routing is point-to-point circuit switching. When a telephone call is being made, a connection is made using a hardwired connection (or series of connections) until a complete circuit is made from origination to destination.

When digital switching was implemented, this principle still did not change. The only difference was that digital electronics were being used in place of electromechanical relays to make the connections. Since digital switches used computers, additional intelligence could be exchanged between the end points using a separate signaling data network.

Packet networks, on the other hand, do not rely on fixed circuits to deliver their payload. The packet network uses a series of connection points to form a network consisting of many different routes. This allows data to be routed through any available connection until it reaches its destination without the need to establish a fixed connection end to end.

This is certainly a more efficient means of networking, but when used for voice and other real-time applications (such as video), it presents some challenges. The biggest challenge is to ensure that there is no delay in the delivery of the packets carrying digitized voice, and to ensure that the packets are delivered in the proper sequence.

Imagine if you be would having a conversation with a family member, through a packet network that is unable to provide this level of quality of service. Your voice packets are routed through various routes, some of them taking longer to reach the final destination (your phone) than others. The end result is scrambled audio that becomes undecipherable.

This is not to mention how delayed packets (latency) causes an undesirable effect similar to satellite connections where the two-way conversation becomes difficult (you begin talking, but suddenly you hear the other party talking). You see this effect on television today when newscasters are broadcasting remotely using IP-based services over the Internet, with no QoS implemented. There is a delay between the local newscaster and the remote newscaster, the video is choppy, and the audio is not in synch with the newscaster's lip movements (the audio and video packets are out of synch).

This has prevented telecommunications networks from evolving to packet networks capable of delivering all forms of communications, until the last several years. This chapter describes the architectural changes and challenges that have led to the biggest changes in telecommunications history: Voice over IP (VoIP).

The Traditional Voice Network

Traditional voice networks use dedicated circuits for connecting end to end for any one call. These circuits remain dedicated through the duration of the call but then remain idle until the next call. This, of course, is a highly inefficient means of utilizing facilities, as they are not being used 100 percent of the time.

Over the years much work has been done to improve trunk utilization; however, the real problem with the entire model is the switching concept that has been in place since the first telephone was connected. As I've already described, the basic model of circuit switching worked well in the early days of telephony but has long outlasted its usefulness.

Packet networks introduce more efficiency, but packet switching also has its issues when dealing with real-time payloads such as video and voice. Packets must be delivered without delay, and in sequence. Packet networks were not designed with real-time applications in mind and therefore were never well suited for voice.

This has all changed now with new protocols developed for session control over packet networks, and the VoIP network engineered specifically for voice and video. But there are still concerns that need to be addressed beyond QoS. It helps first to understand the legacy architectures and why they have remained so reliable and secure when designing a VoIP network.

Wireline Network Architecture

In a wireline network, QoS was guaranteed, since the line feeding from the central office to the residence/business was used for nothing but that service. This is one reason the cost of sustaining a residential subscriber is so high; what is often referred to as the "last mile" is expensive to install and expensive to maintain, with not near enough return to make it profitable.

Yet there are many inherent advantages to this type of implementation, for both QoS and security. Access is considered "secure," since it is a hard-wired connection from the switch to the device, and unless there has been a physical breach, the party who is accessing the circuit and associated services is assumed to be the legitimate subscriber. This is a challenge for VoIP services, where access is obtained through many different technologies, all vulnerable to security breaches.

There are also performance issues in the switched network. Circuits may often sit idle and unused for long periods of time, especially if the network has been engineered poorly. I have spoken with many operators running networks of all sizes who have found entire routes of facilities sitting idle for very lengthy periods of time because of poor engineering practices.

This all comes with a cost as well, so poor engineering of facilities results in loss of profit for the operator. This is one of the reasons IP has become attractive to many operators. The savings it represents in many areas is worth the added effort to make them secure and provide the same QoS as their legacy networks.

Routing within circuit-switched networks is also quite different than in packet networks. The switches in the network are programmed with routes, dictating which circuits will be used for each call type/destination. The routing is based on the digits dialed, and the routes are hierarchical in nature.

This form of routing ensures QoS, because enough circuits can be engineered per route to guarantee available circuits at any time. The unfortunate factor of this routing is that the routes must be engineered for worst-case scenarios (usually the busiest calling day of the year, which is Mother's Day in the U.S.).

Of course this means that the network has been engineered for the heaviest of traffic, which occurs once a year. The rest of the time these circuits sit idle. This is the reason many network operators groom and monitor their routes to ensure there are just enough circuits available to support a typical day of traffic, while engineering contingency routes for heavier days.

Figure 1-1 shows the switching hierarchy model used in the U.S. This is much flatter than models used prior to the divestiture of the Bell System. Prior to divestiture there were five levels of switching. Each level provided an additional route in the event of congestion at the lower levels.

This model was abandoned because of cost postdivestiture. Today's network architecture looks the same worldwide, with two levels of routing. The end office provides the local connection as well as routing within its own region, and the tandem provides connectivity to other regions within the same network. The point of presence (POP) is a U.S. entity providing a gateway function between networks. This could be a function within the tandem switch itself, or a stand-alone function.

Long distance and international calls are then routed to gateways connecting to other networks. These international gateways also provide security and some protection to the network, since the connections must be negotiated and approved by the two connecting operators. Without a connection, there is no access to the switched network.

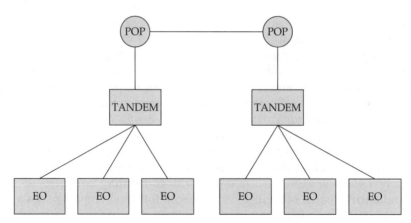

Figure 1-1   Switching hierarchy

Interconnection agreements also dictate the revenues to be paid by each connecting operator when routing a call into another network. Interconnection is a major source of revenue for all operators and is closely managed and monitored. Regardless, there is a fair amount of fraud at the interconnect level, representing a major loss of revenue as well.

This is why many operators are hesitant to deploy IP networks. The reputation of the Internet has impacted the deployment of IP in many traditional service provider networks. It has also spawned a significant amount of work by standards bodies looking to make IP as reliable and secure as their current circuit-switched networks.

The 3rd Generation Partnership Project (3GPP) has released a number of standards that have become known as the IP Multimedia Subsystem (IMS); they define how to implement both IP and SIP to provide traditional services as well as support new 4G wireless services. My book *The IP Multimedia Subsystem: Session Control and Other Network Operations* (McGraw-Hill, 2007) provides more details about this work, so we will not go into detail in this book.

The wireline network also uses databases to support some services. For example, name display is only possible through the use of an IN architecture where databases storing the name and number of all subscribers are kept in the core of the network and accessed by the switches in the network using the SS7 protocol.

The switches send a query to the database containing the telephone number of the calling party, which is then matched in the database to a name. The name is then provided in the response back to the terminating switch for display on the called subscriber's device. This is just one simple example of how services are supported through IN.

Number portability is driving growth in SS7 networks all over the world, because without SS7 and an IN architecture, number portability would not be possible. The concept of number portability is simple; telephone numbers are typically assigned in blocks to each switch in the network. Each switch owns its own block of telephone numbers.

When routing a call in the network, one only needs to know which switch in the network owns the number dialed. Number portability databases identify which switch a particular number has been "ported" to, so the call can be routed to the correct network.

There are many more examples of IN services, but we won't go through them all. What is important is to understand that IN in this context does not refer to the products of various vendors labeled as "IN" or "AIN." The term IN in this context refers to the Intelligent Network architecture that was defined in the 1960s as part of the digital network evolution. The IN architecture relies on SS7 for call control and is analogous to IMS and SIP today.

In fact, early versions of the IN called for SS7 to be used end to end, from the subscriber device all the way to the destination device. Operators at that time felt this was a security risk, since SS7 was the "keys to the kingdom," and therefore SS7 was relegated to the core of the network. The Integrated Digital Services Network (ISDN) was developed as an extension of SS7 to the subscriber premises in response to the security concerns.

This is another important fact to remember, since SIP is analogous with SS7 and the IN, yet today SIP is embedded within the user device rather than just the core of the network. SIP then provides end-to-end session control from the subscriber premises through the network core to the destination device.

There are three major functions of a switch. The *switching fabric* (the matrix that actually completes the connections between each circuit) is responsible for completing a circuit under the control of a computer.

The *call control* portion of the switch directs the connections made in the matrix, and processes signaling information from other switches regarding those connections. It is really the call control function in the switch that provides the intelligence to the switch, and it is what drives the cost of traditional switching equipment so high. Since this function is required in every switch, no matter how big or how small, the cost of deploying a switch can be prohibitive in small markets where there are not many subscribers.

The third function of the switch is the *signaling* function. This is where SS7 messages are processed. This is also the function that manages communications through the SS7 network. In SS7 parlance, this is the service switching point (SSP) function.

We will discuss more about the correlation between these three switching functions and VoIP in a little bit, but before we do, let's look at the various functions of the wireless network and what is required to provide service to wireless subscribers.

The primary thought to take away from this section is that circuits are connected, and then released under call control, and then sit idle waiting for the next call. As you will see, this is very different from VoIP and packet networks.

Wireless Network Architecture

The wireless network is much different than wireline for a number of reasons. In a wireline network, operators are confident that subscribers are who they say they are, because they are attached to the other end of a dedicated cable. In wireless networks, the subscriber accesses the network from any location using a radio device (cellular telephone or other wireless device).

This means that wireless networks must implement other means for authenticating a subscriber, and verifying they are authorized to use the services they are attempting to access. This means using various databases that the network switches can access for these functions.

The wireline network also uses databases, as well as the SS7 protocol to communicate with those databases, but they are not as critical there as they are in wireless networks. Without the IN, roaming is not possible, and wireless service would be very basic.

Figure 1-2 shows a typical GSM network with 2G as well as 3G packet access. The mobile switching center (MSC) provides the connectivity to the various radio sites, acting as a hub in the network. The MSC must access the various databases prior to providing a subscriber services to verify the subscriber is who they say they are (authentication) and to identify the services they are authorized to access.

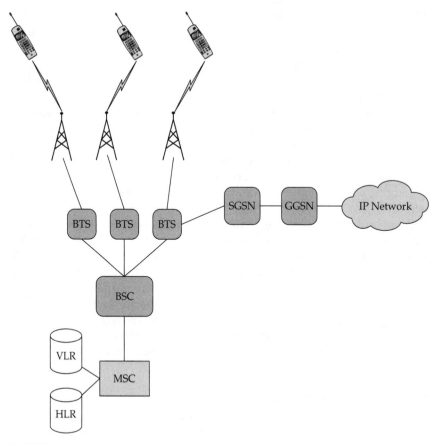

Figure 1-2   GSM wireless network architecture

The MSC also provides connectivity back into the Public Switched Telephone Network (PSTN). This is how fixed-line calls are routed to wireless devices. The MSC also provides connectivity to the home location register (HLR) and the visitor location register (VLR). These two entities provide the functions of the SIP registrar (authentication and authorization).

The HLR provides information about subscribers in their home networks. This information includes their public identities, and the services they are authorized to access. The HLR provides another important function: the location of the subscriber.

The location is identified by first providing the address of the serving MSC. The serving MSC is where the mobile device last registered. The serving MSC address is used to route calls through the network to the device in its current location. The actual cell site serving the mobile device is identified in the VLR. The serving MSC queries its VLR to determine which cell site to route the call through to reach the subscriber.

This concept of mobility works relatively the same as in a SIP network. The SIP network also uses a registrar for registering a subscriber location. It can also provide the authentication and authorization services provided in the wireless network.

The wireless network shown in Figure 1-2 is a simplified illustration meant to provide a basic idea of wireless network architecture. Real networks are, of course, a bit more complex than shown, but this drawing identifies the major components of the network.

Network Elements in a Voice over IP Network

There is one major difference between circuit switching and packet switching; circuits are not dedicated per call as in packet switching. Yes, there are still circuits connecting between each of the involved network entities, but these circuits make up a backbone, if you will, designed to transport millions of calls and data packets all together, rather than selecting a circuit, and reserving that circuit just for one call.

In packet networks, there are still circuits connecting to routers (a router is equivalent to a switch, if you will, in terms of its function). The routers provide the connectivity throughout the network. The circuits themselves are always transporting packets, unlike switched networks, where the trunks or lines sit idle between calls. The packet network is designed to use any available route to deliver all the data and voice packets in the network to their destinations.

This represents a fundamental change in how we route through a network, but it also presents a challenge. Since there are other packets using the same circuits, if the circuits should become congested, packets begin slowing down or are dropped entirely. This has a negative impact on QoS in voice networks, as delay and jitter are not tolerable.

The routers are controlled at the transport layer. Connecting the packet network and providing session control end to end are the equivalent to the call control function within the switch.

As you may recall, in the section "Wireline Network Architecture" we discussed the three major functions of a switch. You will also recall that the call control function was the most expensive part of the switch and the main reason it is so expensive to deploy switches in smaller rural markets.

But what if the call control function could be separated from the rest of the iron? What if the call control function for the entire network could be placed in the core of the network, and smaller, cheaper systems providing nothing more than the switching fabric were deployed at the network edges where service access needs to be?

This is exactly the case with VoIP networks. The switching fabric is a cheaper system that is placed at the network edges for access, under the control of a centralized call controller. The switching matrix is the media gateway (MG), while the control function is the media gateway control function (MGCF). This is shown in Figure 1-3.

The signaling function is inherent in the MGCF, but there still needs to be a gateway function between the legacy SS7 network and the VoIP network, so this is provided by the signaling gateway (SG) (not shown in Figure 1-3). The SG interfaces to the SS7 network using SIGTRAN signaling links (IP-based facilities).

In Figure 1-3 notice the MGCF functions also connect into the same IP network as the MG. The voice and data from various devices all use the same IP network. While the IP network is usually depicted as a cloud, Figure 1-3 illustrates that in fact within the cloud there are routers that are interconnected to other routers. This is not much different than the switched network, except that the circuits are not dedicated to single transmissions in the IP network.

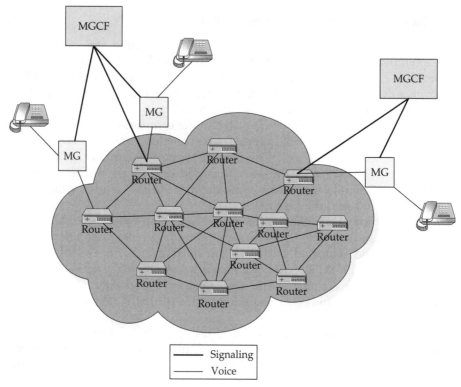

Figure 1-3   A typical VoIP network

So now that you understand the fundamental differences between the various networks, and how databases are used throughout the network for the delivery of common voice services, let's look at each of the SIP network entities specifically and the functions they provide.

Media Gateway (MG)

The *media gateway (MG)* connects the various end points together. In other words, it is the MG that connects a voice facility to the network at the network edges. For example, the MG connects the various circuits out to the customer premises, and converts the payload from these circuits to packet.

Think of the MG as the "packetizer" of the VoIP network. Its job is to take the non-packet circuits, and their payload, and convert them to packet format for transport through the IP network. This means that the other side of the MG supports IP.

This makes the MG rather complex in some terms, since it must support the codecs necessary for converting packetized voice back into audible form for transmission over the voice circuit, and vice versa.

The MG is the equivalent to the switching fabric in the telecom switch. It provides the matrix, if you will, for connecting TDM circuits to packet circuits, and then routing them over the packet network. There will be many TDM circuits connecting to the same packet circuit, however, which is very different from the voice switch.

The MG also provides multimedia connectivity in an all-IP environment. For example, in Figure 1-4 you see how the MG connects via IP to multiple device types at the

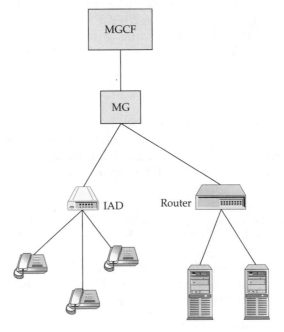

Figure 1-4 MG role in the VoIP architecture

subscriber premises and provides connectivity back into the core network via IP. Note also that the MG connects to the media gateway control function (MGCF). This is where the call control is, communicating with the MG using the SIP protocol.

Other call control protocols could be supported, depending on the implementation, but given our focus in this book is SIP, we will pretend there are no others. SIP in these examples is the call/session control for the entire network.

Media Gateway Control Function (MGCF)

The *media gateway control function (MGCF)* is the most important entity in the whole network, for it is here that all the decisions are made regarding the management of a call. This is the brains of the network, the equivalent to the call control function in the voice switch.

The MGCF determines how a call is to be routed through the network based on signaling from the MG. The MG is responsible for the actual connections, while the MGCF manages the resources of the MG. This means determining what codecs are needed for the session, and signaling to the destination MGCF/MG regarding the session.

The MGCF generates the requests/responses in the SIP network. The MG originates connections and notifies the MGCF using either SIP or some other signaling protocol such as the Media Gateway Control Protocol (MGCP), or H.323. The MGCP also possesses a SIP URI address for routing purposes.

While the MGCF does initiate requests and responses, the MGCF does not register in the network. Its purpose is that of stateful proxy within the SIP domain. The MGCF communicates with other MGCFs within its own network, but it interfaces with the breakout gateway control function (BGCF) when interfacing to other networks. The BGCF acts as a network gateway and can also be implemented as the gateway into the PSTN if the PSTN is another domain (otherwise, the MGCF traditionally provides this interface).

Signaling Gateway (SG)

The *signaling gateway (SG)* interfaces between the SS7 network and the SIP network. It can also interface to other non-SIP VoIP networks such as H.323 or MEGACO. The signaling gateway provides an SS7 ISUP interface into the VoIP network by transporting ISUP over an IP facility. The transport protocol used at the signaling gateway is SIGTRAN.

The signaling gateway uses traditional SS7 interfaces (using the Message Transfer Part, or MTP, for transport) within the SS7 domain, then SIGTRAN into the SIP domain. Many BGCF/MGCFs support ISUP as well as SIP, providing the necessary gateway function.

The signaling gateway is not a SIP entity per se, since it typically does not support SIP. Its primary purpose is to provide an IP interface for ISUP into the VoIP network. The signaling gateway provides an important function when implementing a SIP network.

Application Servers (ASs)

The *application server (AS)* is a SIP entity providing services to the network as well as subscribers. The IN used a similar concept, with servers (or "big iron" processors) providing services such as number portability and calling name display.

Many operators have grown tired of "big iron" requirements for service platforms, so the concept of an "off-the-shelf" server supporting traditional services as well as 3G/4G services is much more appealing.

The application server sits in the core of the network (although this is not a requirement, as the AS can be deployed anywhere in the network). It is capable of generating requests and initiating dialogs acting as a user agent (both UAS and UAC). The AS also has a SIP URI address.

The AS also uses service identifiers to address the various services being supported. This allows the network to send a request to the AS that may be supporting multiple services and direct the AS to provide a specific service.

The registrar function may contain not only the public and private identities of a subscriber; it may also associate service identities with each identity for a subscriber. This allows the registrar to provide authorization as well as authentication.

Some services may require the AS to support event notification. The AS will send the SIP request *SUBSCRIBE* to the appropriate registrar in the network so that it can be notified of any changes to a subscriber's status (change of location, for example, or availability of the subscriber). Services such as presence or location-based services would require event notification. This is discussed in more detail in subsequent chapters.

If the AS is acting as a SIP proxy, requests are routed to the AS, which then will forward the request to the next hop in the network after implementing whatever service the AS is delivering. The AS address may also be included within the request-URI header if strict routing is being used, in which case the AS will remove its address prior to forwarding the request on.

Much of the way the AS behaves is not much different than the service control point (SCP) function defined in SS7/IN networks. The switches in the IN query servers to determine how to deliver services as well. The difference is the ability to support multiple services and multiple terminals on the same platform in the same network.

The Domain Name System (DNS)

The DNS is what provides SIP networks (and all other IP networks) the ability to support mobility. In the Internet, the DNS is used to provide routing addresses for servers providing Web services in the network, as well as find the physical addresses of subscribers for the delivery of e-mail and other media. This requires converting names to numeric IP addresses.

The DNS stores the URI of the subscriber and all other entities in the network with an address. This includes application servers. The URI is then translated into an IP address for routing to the final destination. This means that the registrar in the network must update the DNS when any user agent registers with a new IP address.

When the user agent registers, it provides its IP address that has been assigned by the network (which is usually the case when dynamic addressing is used). The DNS must be queried by the proxies in the network when they receive a message to be routed and only the URI is provided.

To prevent overloading the DNS, the function is distributed throughout the network. This prevents the possibility of too many requests flooding the DNS and causing it to fail, thereby denying service to all those whose address is stored in the DNS.

In fact the DNS architecture is quite sophisticated. There are many different levels of DNS, enabling a tiered approach to finding information. This avoids the need to have one central server for all queries.

In fact some servers may not contain the information needed to route to a specific subscriber but know the address of the DNS server that would have the information. The DNS queried would then provide the address of another DNS server to be queried.

There may be multiple queries to the DNS for a single session, since the address information may not be forwarded between networks. This means that many queries could be possible for a single call. The DNS is the busiest database in the world, receiving millions of queries each day, and changing entries by the millions each day.

To better manage DNS entries, domain names are divided into domains such as .com, .org, and .edu, to name just a few. There is a DNS server responsible for each of these domains. All of the domain names are managed by a third party that administers domain names to recipients all over the world. They are then entered into the domain name servers.

The authority for managing these domain names, as well for updating the master database (referred to as "root"), rests with registrars from other networks. The top level (or root) database then feeds to all other databases worldwide. Notice in Figure 1-5 that there is a DNS for the UK, as well as for each domain (.com, .org, etc.). Some countries choose to maintain their own top-level database, providing addresses for all domains within their country. For addresses outside of the UK, the root DNS is accessed.

To further prevent congestion, corporations may also maintain their own DNSs. This eliminates the need to access the DNS of an external network. These DNSs then connect to an external DNS within the hierarchy until the address is "resolved" to an IP address. For example, an e-mail sent to travisruss@tcg.com could be sent to the DNS maintained by my ISP.

If my ISP cannot resolve the address, it would then provide the IP address of the .com DNS, which could then resolve the address. This is an oversimplified example of how the DNS works and is not meant to be an exhaustive study of DNS servers, since that is outside the scope of this book. There are plenty of tutorials on the Internet that provide exhaustive information about the workings of the DNS.

Electronic Numbering (ENUM)

When VoIP came along, the need to support telephone numbers challenged the DNS. As discussed previously, the DNS resolves names into IP addresses. Telephone numbers are not included as names; therefore, another database is needed to support resolving telephone numbers to IP addresses.

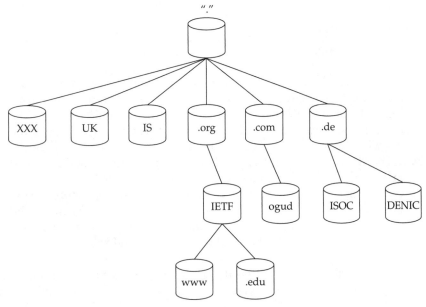

Figure 1-5   Domain name server hierarchy

There are many thoughts regarding the best approach to ENUM. Some believe that maintaining local ENUM servers (each network would have their own) is the best solution, while others believe the DNS and ENUM should be colocated within the same server.

Regardless of how ENUM is implemented, the purpose is to provide an IP address for each telephone number in the same manner the DNS resolves URIs to IP addresses.

SIP-Specific Entities

So far we have discussed elements commonly found within VoIP networks. There are other entities that are specific to SIP networks, which we will discuss next. Some of these entities are found in other network types such as the Internet. This is because SIP was derived from many functions found on the Internet now. In fact the routing, addressing, and session management are already implemented in the public Internet.

The SIP protocol actually came from the Simple Mail Transport Protocol (SMTP) and the Hypertext Transport Protocol (HTTP). This means many of the functions we will be talking about in later chapters also came from these networks. SIP was developed to provide five important capabilities that modern-day networks cannot achieve:

- Location of the participants in a session
- Participant availability

- Participant capability
- Session setup
- Session management

To achieve each of these capabilities, SIP requires a network that can provide specific functions. It also requires specific functions within the network entities and user devices to be able to achieve all of the preceding capabilities.

Location of participants is achieved through the registration of a user (or user agent, as you will learn in the next section). The registration process is what allows SIP to learn the IP address of any subscriber when that subscriber connects to the network.

Participant availability is about more than whether their device is turned on or off. The subscriber may have options to forward calls to voicemail, for example, or to only receive specific types of calls. The subscriber may create profiles that determine how calls will be routed to that subscriber during specific times of day. Application servers are used to store the availability of a subscriber and to notify querying entities of the subscriber's availability.

Each subscriber's capability is determined by the device he or she is using. It is up to the user agent to determine what capabilities the device has and to communicate that to the other user agents.

SIP then creates a session using specific message types (or methods, as they are referred to in the RFCs) and managing those sessions. We will be talking about session establishment and management throughout this book. We will also define each of the entities from a SIP perspective to facilitate later discussions about SIP routing and networking.

User Agents (UAs)

A *user agent* is software embedded within a device that is capable of generating requests or sending a response to a request. The user agent resides within subscriber devices, proxies, and application servers.

It is the user agents that react to one another. When a dialog is created, it is a user agent communicating with another user agent. There are two types of user agents, both residing within the same devices but providing different functions: the user agent client (UAC) and the user agent server (UAS).

User Agent Clients (UACs)   Only a UAC can originate a request. Therefore any device or network entity that needs to originate requests must have this function. Simply put, the UAC is the originator of all sessions. The client establishes a dialog and maintains that dialog with the user agent at the destination.

The UAC coexists with the UAS. In other words, a device will have both functions. It is also possible to have a session where a device is playing both roles; UAC and UAS for the same session. This happens when there are session modifications. For example, the destination device may wish to add another participant to a call in progress.

The destination device would then generate a request (using the UAC function) and would maintain the state of that request throughout the dialog.

The UAC generates a request (the beginnings of a session establishment) when the subscriber dials his or her phone, or uses some other application for communicating with the network. This includes e-mail, instant messaging, text messaging, and accessing the Web. The UAC interprets the commands received from these applications, including the addresses input by the subscriber and/or the application.

The addresses may be simple words (such as "Travis" selected from an address book). The UAC must then determine what this address is and how to route to this address. The application (address book in this case) would provide more details behind the address, such as travis.russell@tcg.com, but the UAC must resolve this address to an IP address.

This is done prior to routing into the network. Therefore the UAC must make a lot of decisions regarding the treatment of a session and how it is going to be routed through the network. We will be discussing much more about the role of the UAC throughout this book.

User Agent Servers (UASs)   The UAS responds to requests from other UACs. Think of the UAS as the recipient of a request. The UAS must then send responses (either $1xx/2xx$, or failure responses). The UAS can also be a stand-alone function; some devices may not have the UAC capability but do have UAS capability. Some application servers, for example, may have the ability to respond to a request but be unable to generate a request.

The UAS may respond with capabilities of the device, and of course whether or not it accepts a session invitation. The UAS may act independently or in conjunction with device applications. For example, if a subscriber is being asked to join in an instant messaging session, the instant messaging application may allow the subscriber to accept or reject the invitation. The UAS would then communicate the acceptance or rejection back to the UAC that originated the request.

When a request is received by the UAS, it first determines if the method being sent is supported. For example, if the method *MESSAGE* is sent, the UAS must determine if the device supports messaging and therefore supports this method. If the UAS determines it cannot support the method, it will send a response along with parameters identifying the methods it does support. This is part of the capability discovery.

The UAS must also determine if there are headers contained within the request that it does not understand. If there are headers it does not understand, the message is not rejected but processed, with the exception of the headers the UAS cannot process. This is necessary because proxies are also considered user agents, and therefore there are many parameters within a request not intended for a proxy that it would not understand.

There are a minimum set of headers that any UAS must be able to understand and process, to ensure that all devices will work in the SIP domain. This allows devices to be added to the network even if they do not support proprietary functions (and therefore proprietary SIP extensions).

Vendors who build SIP devices may also introduce SIP methods and headers that are understood only by their software. This could be a problem when another vendor's equipment is introduced into the network (such as a proxy, or application server). For this reason, the UAS will ignore methods and headers it does not understand and process as much of the message as possible.

Since all devices support a minimum set of SIP methods and parameters, all devices are guaranteed the ability to communicate with one another and generate requests and responses. This is a unique capability of SIP and its user agents not found in many other types of networks.

Proxy Servers

A *proxy* receives SIP requests and responses and forwards them through the network on behalf of the user agent. Think of proxies as routers that are also capable of generating requests and responses. You will learn later about how proxies will send certain responses backward to the UAC, for example, to notify the UAC that it is forwarding a request on to the recipient UAS.

The proxy is also responsible for determining how to route requests and responses. This can be done through two methods: loose routing and strict routing. We won't go into the specifics of either one, as they are described in detail in Chapter 5.

Next we will discuss several different types of proxies: stateful proxies, stateless proxies, and forking proxies.

Proxies should be considered as routers in the network. But routers are only concerned about forwarding packets to the next hop in the path. They only know about the other routers in the network and do not concern themselves with the status of the session itself.

This is where SIP proxies are different. The SIP proxy must be able to resolve URIs like a router, but the SIP proxy must also be able to send responses when a request is received. The SIP proxy is also equipped as a user agent, capable of requests and responses.

Stateful Proxies   A *stateful* proxy must maintain the state of each and every dialog that it is part of. As a request is sent to the proxy, it saves the call identifier for the session, as well as other pertinent identification data. This means that in order for the stateful proxy to know when a session has changed status or has ended, it must receive all responses.

This is why networks that use stateful proxies must route all messages for a session through the same route. This ensures that all SIP responses use the same path as the request, so that the stateful proxy will receive all of the responses and any additional requests for a session. Each stateful proxy adds a new *VIA* header with its own address as the topmost *VIA* header to maintain routing order. These are then used in reverse order for routing responses.

The *VIA* header is used to ensure that the responses are sent through the same proxies as the request, or routing can be forced through the same path using strict routing methods. In addition to maintaining the state of the session, there are several other functions of the stateful proxy.

A stateful proxy must process all routing data to determine the target(s) for the request. It must then determine how to route the request to the destination and process all responses from the recipient(s). It will also send a 1xx response to the UAC to prevent retransmission of the request and to let the UAC know that it has received the request and is forwarding the request to the next hop. This is where the proxy behaves as a UAS. The stateful proxy cannot send 2xx responses.

The stateful proxy must also perform a series of checks on requests and responses to ensure they meet the proper criteria. This includes checking the syntax of a message to ensure it is correct prior to forwarding to the next hop. If the syntax is not correct, the UAS of the stateful proxy will return an error response.

Likewise, the stateful proxy will also check the URI provided in the request to ensure that the URI format can be understood by the proxy. If the URI is not of the correct format, the message cannot be forwarded, because the proxy does not know where to send the message. Again, the proxy will send an error response back to the UAC.

The proxy also looks at the *MAX-FORWARDS* value to ensure that the request has not already been through too many hops. This would indicate that the message has been through too many networks or is possibly being sent in circular fashion. The proxy will also look to see if its own address has been included in the *VIA* header, indicating that the request has already been through the proxy and the request is being circular-routed.

When a proxy adds its own address to a *VIA* header, it computes a *BRANCH* value by taking the tags from the *TO* and *FROM* headers, along with the values from the *CALL-ID* and *REQUEST-URI* headers, the topmost *VIA* header, and the *CSeq* header and creating a cryptographic hash from all of these values. This then becomes the latter portion of the *BRANCH* value. The stateful proxy then checks for this value whenever a request is received and the proxy's own address is already present in a *VIA* header.

If there are headers in the request or response that the proxy does not understand, it will not reject the message. This is because there are many extensions to the SIP protocol that have not been defined by the IETF but are proprietary extensions added by vendors. This makes them proprietary and therefore unknown to the stateful proxy. For this reason they are ignored to ensure interoperability between all SIP entities.

A stateful proxy can terminate a session by using the *CANCEL* method if no 2xx response is received. Each stateful proxy possesses a timer that is set upon the receipt of a request. If that timer expires prior to the 2xx being received (or any other final response), then the stateful proxy will respond to the UAC with the *CANCEL* method to terminate the session.

Stateless Proxies   A *stateless* proxy does not have any concept of a dialog or any knowledge of the state for any dialog or session. The stateless proxy is nothing more than a simple router responsible for nothing more than the forwarding of a request or response. It does not maintain any knowledge of previous requests or responses, and it does not maintain any routing data other than a routing table of its own used to determine how to route messages based on the URI provided in the SIP message.

This means such a proxy cannot respond with any provisional 1xx responses, nor can it terminate any sessions should a final response get lost or fail to be generated. The stateless proxy does compute a *BRANCH* value to be appended in the *VIA* header in the same fashion as the stateful proxy for identifying looping in the network.

Forking Proxies   *Forking* proxies are used to split requests for routing to multiple destinations. For example, if a subscriber has registered multiple addresses (for multiple devices) and has a service that facilitates routing of calls to all devices, a forking proxy would then take the original request and split (or fork) it into multiple requests.

These multiple requests would then be sent to all of the destinations registered for the subscriber. This is only one example; there are many other, similar services where forking of requests requires the services of a forking proxy.

Since the forking proxy is splitting the request, it is the only entity that knows about the other destinations. The forking proxy may be an application server providing proxy service. Whatever the implementation, the originator of the request has no knowledge of the request being sent to other destinations. This means that the forking proxy must manage the responses from the other destinations.

This of course requires the forking proxy to be a stateful proxy. As a stateful proxy, the forking proxy will send a 1xx provisional response when it receives the request. It will then forward the request to multiple destinations. However, when it receives multiple responses, it does not forward those responses to the UAC.

This is one important differentiator between forking proxies and stateful proxies. Stateful proxies do not send 2xx responses, but the forking proxy does if it receives at least one 2xx response from one destination. The forking proxy will send the 2xx response back to the UAC regardless of the responses sent by the other destinations.

This means that any non-1xx/2xx responses (error responses) must be handled by the forking proxy and not sent to the UAC, as this will cause the session to be terminated. Therefore, the proxy will return the appropriate response to all other destinations, depending on their response.

For example, if multiple destinations send a 2xx response, the forking proxy will return an *ACK* to each of the responding destinations (remember only the forking proxy knows who these destinations are). The forking proxy must then also maintain the dialog between itself and the other destinations, acting as a middle agent for the entire session.

Likewise, if any of the destinations responds with a 3xx–6xx response, the forking proxy will return a *BYE* message and terminate the dialog between itself and the other destination. The UAC that originated the request has no knowledge of this and has no visibility into the responses.

Redirect Servers

A *redirect* server is used to provide alternate addresses for a request. There are many reasons why this would be desirable. For example, the network operator may wish to send alternate addresses for routing of a request when proxies become busy as a means of load sharing.

The redirect server sends a 3*xx* response when it receives a request. The 3*xx* response provides the alternate address(es) to be used to reach the destination. The addresses provided could come from a variety of sources. One example would be a location server providing addresses based on current registrations to the redirect server upon query.

The redirect server would then use these addresses in its own 3*xx* response back to the UAC, which would then send the request directly to the address(es) received. This definitely helps in reducing the amount of processing required on the proxies in the network.

Registrars

SIP registrars are used to authenticate and record the current location of a subscriber device. When a device is turned on or changes location (resulting in a new IP address being assigned), the device will send a REGISTER message into the SIP network to provide its new address.

The registrar then has two options: One is to accept the new address and save it in a location server. The other option is to challenge the subscriber by rejecting the first registration attempt and force the subscriber to send authentication keys (credentials) to verify they are indeed who they say they are.

This second option is highly recommended in any SIP network as a very simple security measure that helps prevent man-in-the-middle attacks. Forcing authentication as part of the registration process would eliminate a lot of fraud and security issues today if it were implemented; sadly, there are many VoIP implementations that do not force authentication or authorization at all.

Location Servers

Location servers provide subscriber addresses to other proxies based on the registration process. This function is not necessarily a stand-alone function but rather integrated within a SIP server providing registrar and proxy services.

The location server gets its data from multiple sources, including the DNS. It could also get information from home location registers (HLRs) and visitor location registers (VLRs) in the wireless domain. Of course within the SIP domain the primary source is the *REGISTER* method to the registrar.

The data provided by the location server can include such things as availability, service profiles, SIP URIs, GPS coordinates, and preferences as well. There are also a number of advanced functions that are being defined for location servers, such as those described by the 3GPP for IMS implementations.

SIP is not used to query the location server. This is handled by another protocol altogether and is outside the scope of this book. RFC 3263 provides more details as to the procedures used to access and query location servers.

2

Structure of the SIP Protocol

When Voice over IP (VoIP) was first implemented by some of the pioneering service providers, there were many challenges they faced in managing voice communications in an environment primarily developed for data. The challenge comes in the difference between data and voice communications.

In a data environment, delivery is not expected in real time. Nor does delivery need to be guaranteed in many cases. For example, an electronic mail can be sent, but delivery to the end user may take days simply because the recipient is not available (requiring the e-mail to be stored on a network server and forwarded when the recipient requests). Another example may be where the network is not able to deliver the e-mail immediately due to other traffic or congestion, in which case there is a latency issue.

Instant messaging is another good example where a conversation is in place, albeit the conversation is in text form. A delay between responses is acceptable given the nature of instant messaging and the culture behind it. If a reply is not received for several minutes, the conversation does not end.

SIP Messages and Formats

Voice communications by pure nature is a real-time application. Delays in the delivery of voice packets are not tolerated because the end result is long periods of silence between the two parties during the conversation. We expect an experience that would suggest a full duplex communications between two parties, but with conventional data protocols this becomes a challenge.

For this reason the Internet Engineering Task Force (IETF) began defining a protocol designed specifically for the control of real-time multimedia communications. The intent was not to limit the requirements to support voice, but to create a session control protocol capable of supporting all forms of communications, regardless of the media type.

There have been many different versions of call control protocols developed for this purpose since the inception of VoIP, but SIP became the favorite because of its ability to support multiple media types. SIP is not really all that new, however.

SIP was derived from the Hypertext Transport Protocol (HTTP), so you will see many aspects of the SIP protocol that resembles HTTP. It is text-based, which makes it simpler to understand than most bit-oriented protocols. In a bit-oriented protocol, you must know the significance of each bit position according to the rules and syntax of the defined protocol.

There are also some aspects of SIP derived from the Simple Mail Transport Protocol (SMTP). You will see aspects of this in the SIP headers and parameters described in this chapter, as well as in the routing of the SIP messaging through the network.

It is important to understand that a SIP entity is not a physical "thing." SIP entities are logical. Indeed, a physical entity such as a server may support several logical entities. Each SIP network entity must be capable of processing SIP messages in stages, referred to as layers. Not every SIP entity is capable of all of the layers; depending on the entities functions, there may be no need for some of the layers. All SIP entities must be able to support the first two layers. The layers are defined as:

- Syntax and encoding
- Transport layer
- Transaction layer
- Transaction user

Syntax and encoding is the lowest layer defined in the SIP protocol. Syntax and encoding is essential to understanding and interpreting or processing any SIP message, so this is mandatory for all network entities. This is best defined as the set of rules that defines and constitutes the format and structure of each of the SIP messages. In short, it allows the entity to understand and interpret SIP messages.

For example, when a SIP message is received by a network entity, it needs to know the difference between the various message types, and where the parameters begin and end for each message header. This is part of the syntax and encoding responsibility.

The transport function is a very basic process within every entity that supports the need to create a message and send it through the network to another entity. All SIP entities have a transport layer. The transport layer defines how communications between two elements will be supported using the TCP or SCTP protocols. When a connection is established between two entities, the transport layer creates an index to the connection using the IP address, the port number, and the transport protocol.

When the connection is made, the index consists of the destination IP address, port, and transport protocol. When the transport layer is accepting a connection, the index consists of the source IP address, port, and transport protocol. We will talk more about the management of communications between two entities when we discuss the basic concepts of a dialog.

The transaction layer is a fundamental part of SIP. A *transaction* is defined as a request sent from a SIP client to a SIP server using the transport layer, including all associated responses to the request. The transaction layer is responsible for managing retransmissions, correlating responses to their associated requests, and timeouts. Only user agents and stateful proxies have a transaction layer.

All SIP entities that create SIP transactions are considered transaction users. A transaction user is an entity that creates SIP messages such as an *INVITE* and also has the ability to cancel transactions. The transaction user is also responsible for identifying the addressing for each transaction. It will include the destination IP address and port, as well as transport along with the transaction request.

All communications in a SIP network require a transaction. For example, if a subscriber is establishing a connection with another subscriber, a transaction is initiated. If the session is canceled for any reason, SIP treats this as a separate transaction (although it will be associated with the initial transaction used to establish the session). A subscriber may initiate many different sessions, requiring many different SIP transactions in the network. Each transaction is considered separate and unique by the network elements, yet they can also be associated with one another.

A transaction is not the same as a dialog. A dialog is established only after a sequence of transactions has been completed. This is necessary for a number of reasons, which we will discuss in the next section.

Concept of a Dialog

When two devices communicate with one another, they exchange a set of messages (referred to as transactions). For example, a cell phone originating a call to another cell phone would send an *INVITE* to the other phone requesting a connection be made. The recipient of the *INVITE* will determine whether to accept or reject the transaction.

If the cell phone chooses to accept the invitation to a session, then it will send a response to the request and exchange other messages, entering into a dialog with the other device. The dialog then becomes a logical connection between the communicating entities for the purpose of exchanging SIP messages regarding a session.

Each user agent establishes its own dialog with the user agent client (the requestor). The dialog is then used by the user agent to maintain the status of the dialog and associated sessions. The session is not the same as a dialog, since a session can involve multiple user agents communicating with one user agent client.

In other words, the dialog is established between each entity involved in the same session as depicted in Figure 2-1. The illustration shows the user agent client with a dialog established with multiple user agent servers, exchanging session control information for the same session (as would be the case in a conference call).

As can be seen in the illustration, the user agent client can now delineate transactions between each of the entities participating in the same session. This allows the UAC to correlate responses from each individual UAS and treat each one independently even though they are all participating in the same session.

Each dialog requires a dialog ID, which is derived from the SIP headers. When the UAC sends a request for a session, it will expect a response. In that response will be required headers (as we will discuss in a moment). The response must contain the *TO, FROM,* and *CALL-ID* headers. The *TO* and the *FROM* headers will include the *TAG* parameter. The *TAG* parameter is used by each user agent for correlating requests with responses, but it is also used for calculating the dialog ID.

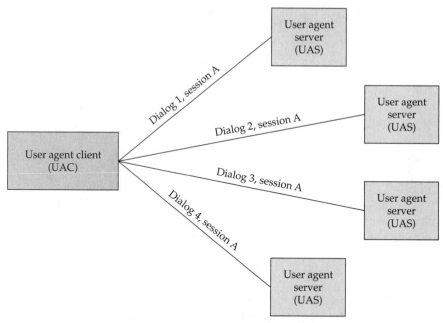

Figure 2.1 The difference between a dialog and a session

The dialog ID then becomes unique for the UAC and each of the UASs. In other words, the UAC creates its own dialog ID, while each of the UASs will create its own unique dialog ID. This is not communicated to other entities, since the use of the dialog ID is local (used by the user agent).

Note that a dialog ID cannot be determined until the UAC receives a *2xx* response back from each of the UASs. This is because the *TO* header does not actually contain the *TAG* parameter until the response is generated by the UAS. Remember that the *TAG* is used by the UAS for correlation of responses (more on that when we discuss the various message headers and their parameters).

So in summary, a UAC will generate a SIP message and send it to one or more UASs. Each UAS will then answer the UAC by sending a response message. The response message contains at a minimum the *TO, FROM,* and *CALL-ID* headers, which are used in creating the dialog ID. Once all of these are done, the dialog is considered established, as depicted in the call flow shown in Figure 2-2.

Figure 2-2 also shows a handshake. Once the dialog has been established and the acknowledgment is received, the handshake sequence between two entities has been completed. This marks the beginning of a session where bearer traffic is exchanged. A session cannot begin without completion of a handshake between the two entities. Think of the handshake as an agreement between all involved user agents to engage in the transfer of bearer traffic.

To establish a session, a sequence of messages must be exchanged between all involved parties. This sequence of messages consists of an offer and an answer. The offer is established by the user agent client wishing to begin a session. It must contain all of the data necessary to establish the session and transfer bearer traffic.

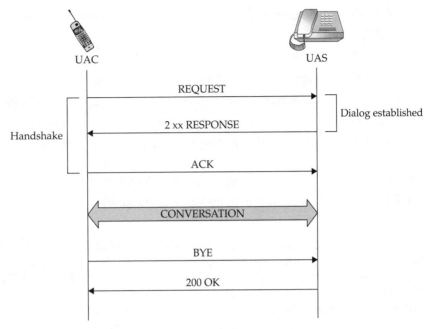

Figure 2.2   Establishment of a dialog

The offer therefore carries the description of the session, including the encoding methods, required media support, and any other support required for the session at the receiving end. The offer may or may not be contained in the initial request, but carried in a subsequent response.

The offer is then followed by an answer. Each entity involved in the session must provide an answer to the offer. The answer in essence is accepting the offer and acknowledging it can support the requirements specified in the offer. If no requirements were specified in the offer, the answer may contain requirements.

The offer and the answer use specific methods. A method in SIP parlance is a message type. There are two forms of SIP messaging: the SIP request and the SIP response.

Requests

A request is sent from a SIP client to a SIP server. A SIP server could support any type of function. Remember these are logical entities rather than physical entities. The request format consists of a start line containing the following information:

```
METHOD (space) REQUEST URI   (space) SIP VERSION    (crlf)
```

The *METHOD* identifies the type of request being sent. There are several different methods supported. There are sometimes other methods defined by various vendors, referred to as extensions. Those are discussed later. The standard SIP methods are as follows:

- *REGISTER*   Used to register a user
- *INVITE*   Used to invite a user to a session

- *ACK*   Acknowledges receipt of a request when setting up a session
- *CANCEL*   Cancels a transaction
- *BYE*   Terminates a session or transaction
- *OPTIONS*   Queries servers about their capabilities
- *INFO*   Used to exchange mid-call information such as dialed digits
- *MESSAGE*   Used for the support of short message service and instant messaging
- *NOTIFY*   Used to notify entities subscribed to event notification of registration updates
- *SUBSCRIBE*   Used to subscribe to event notification
- *UPDATE*   Allows a device to update session information during a session without sending a re-invite

The *REQUEST-URI* is explained later in this chapter, in the section "SIP Identities." Addresses in a SIP network are referred to as Universal Resource Identifiers (URIs). These are very much like Universal Resource Locators (URLs) used in the HTTP protocol (the Web addresses we have come so familiar with). The *REQUEST-URI* is used for routing the SIP message through the network, so therefore it contains the address of the next hop.

Last but not least in the request message is the *SIP VERSION*. This identifies the version of SIP that is being supported by the transaction user that created the SIP message. This allows the receiving entity of a request to understand what version of SIP was used to create the request, so it can decode the request properly.

A typical request start line may look like

```
INVITE sip:russell@tekelec.com SIP/2.0
```

Following the start line in a request are the message header (or headers), a carriage return line feed (CRLF), and an optional message body. Header fields and the message body are described in a later section.

So the request identifies the type of session that is being requested by the user agent client. The requirements for supporting the session (such as the type of media, how it is encoded, and what other methods must be supported) are included as part of the request. There are requests that are specific to certain types of sessions. A good example of this is the *MESSAGE* method, which carries a text message. No session and dialog are needed, since the contents are provided in the message body. The recipient can choose to accept the message or not through a response.

Responses

Responses are sent by a SIP server to a client in response to a request. Responses are identified by the receiving entity by the status line, which is formatted differently than a request start line. The response status line contains the following information:

```
SIP VERSION    (space)STATUS CODE    (space)REASON PHRASE (crlf)
```

Think of the response status line as informing the user of the request status. The *STATUS CODE* is a numerical code used by the receiver to identify the status of a request. It is a three-digit code followed by a text field that provides a textual description of the code for us humans that are number-challenged. While the various network entities will be able to interpret the numeric code, there is no reason as a technician or engineer to memorize these codes, since they are always followed by the *REASON PHRASE,* which is always in plain English.

The status codes are defined by class, the first digit indicating the class of the code. There are currently six classes defined:

- 1*xx*   Provisional
- 2*xx*   Success
- 3*xx*   Redirection
- 4*xx*   Client Error
- 5*xx*   Server Error
- 6*xx*   Global Failure

There are processes that are associated with each of these classes, explained in subsequent chapters. The provisional class indicates that a request has been received and is being processed. This is a normal response to a request, serving as an acknowledgment. The purpose is to prevent retransmission of the request, as we will discuss later. Provisional status codes are defined in more detail in Chapter 3.

Success indicates that the request has been processed and action has been completed. This response would follow a request requiring some form of action to be taken by a server.

Redirection is sent when a session is being redirected to another address for the same subscriber. For example, a request may be sent to a subscriber address, but the subscriber has provisioned his or her services to redirect all requests to another device. The redirection response will indicate the reason for the redirection and provide additional details as to where the message is being redirected.

Client error indicates that the requestor (the SIP client) sent a request with bad syntax or some other form of error, and the SIP server is unable to process the message. Server error status codes indicate an error on the part of the message recipient, or the SIP server. The request itself was valid, but for the reason indicated in the error message, the server was unable to process the request.

Finally, global failure status codes indicate that a request cannot be fulfilled by any SIP server for the reason indicated by the individual code itself. The exact wording for each of the status codes is defined in the standards; however, there are provisions made to allow for other languages. While the standards highly recommend the use of the defined text, the defined text may be replaced by other descriptors such as a different language.

Now that you understand the format of a SIP message, let's look at the various headers and how they are used. The next session will describe the headers as defined by the IETF and how they are used in a SIP network.

Header Fields

Header fields contain detailed information about the request or the response. This includes the destination and origination addresses of some requests, as well as routing information. The header field uses the form of

```
header name: header value
```

There can be spaces between the header name and the header value fields, so you may encounter implementations of this; however, the standards highly recommend against the use of white space between the header name and the header value, recommending the use of a single space between the two.

There can be multiple appearances of the same header name within one request, such as the *ROUTE* header name. When this appears multiple times, the order is not specific, but it is recommended that routing should be listed in the top of the request so that the routers and proxies in the network are able to parse these messages quickly.

There can be mixed orders, though, for example:

```
ROUTE: <russell@tekelec.com>
ROUTE: <jones@tekelec.com>
SUBJECT: Lunch at 1:00
ROUTE: <smith@tekelec.com>
```

The point is, while the standards do recommend using some form of order in the headers, it is not required, and the header names can be in a mixed order. The receiving entities should still be able to process these requests even though the order is mixed (although it may slightly slow the processing of the request).

Another worthy note is that the contents of the header field are not case sensitive. The exception to this rule is when a value is part of a quotation. This would be the case when text, for example, is to be displayed on a subscriber device. In this event, the text is displayed exactly as it is found within the header field.

There are two classifications for header fields: request header fields and response header fields. Some header fields only make sense in a request, for example, and therefore are ignored if they are received as part of a response.

Header fields can be represented in both long form and abbreviated form. The RFC has defined short, abbreviated versions of each of the header fields for use where bandwidth is a concern. The abbreviated versions are provided with each of the header field descriptions that follow.

Table 2-1 identifies all of the possible header names, and what methods they appear in. The *INFO* method has been left out of this table because no information regarding actual required headers could be found.

TABLE 2.1 Header Field Summary

Header Field	ACK	BYE	CAN	INV	OPT	REG	MSG	NOT	UPD	SUB
Accept		X		X	X	X	X	X	X	X
Accept encoding		X		X	X	X		X	X	X
Accept language		X		X	X	X		X	X	X
Alert info		X		X	X	X	X	X	X	X
Allow		X		X	X	X				
Authentication info				X						
Authorization	X	X	X	X	X	X	X	X	X	X
Call ID	X	X	X	X	X	X	X	X	X	X
Call info				X	X	X	X		X	
Contact	X	X		X	X	X	X	X	X	X
Content disposition	X	X		X	X	X		X	X	X
Content encoding	X	X		X	X	X	X	X	X	X
Content language	X	X		X	X	X	X	X	X	X
Content length	X	X	X	X	X	X	X	X	X	X
Content type	X	X		X	X	X	X	X	X	X
CSeq	X	X	X	X	X	X	X	X	X	X
Date	X	X	X	X	X	X	X	X	X	X
Error info		X	X	X	X	X				
Expires				X		X	X			X
From	X	X	X	X	X	X	X	X	X	X
In reply to				X			X			
Max forwards	X	X	X	X	X	X	X	X	X	X
MIME version	X	X		X	X	X	X	X	X	X
Organization				X	X	X	X		X	X
Priority				X			X			
Proxy authenticate		X	X	X	X	X				
Proxy authorization	X	X		X	X	X	X	X	X	X
Proxy require	X	X	X	X	X	X	X	X	X	X
Record route	X	X		X	X		X	X	X	X
Reply to				X			X			
Require		X		X	X	X	X	X	X	X
Retry after		X	X	X	X	X				
Route	X	X	X	X	X	X	X	X	X	X
Server		X	X	X	X	X				
Subject				X				X		

(Continued)

TABLE 2.1   Header Field Summary (*continued*)

Header Field	ACK	BYE	CAN	INV	OPT	REG	MSG	NOT	UPD	SUB
Supported		X	X	X	X	X	X	X	X	X
Timestamp	X	X	X	X	X	X	X	X	X	X
To	X	X	X	X	X	X	X	X	X	X
Unsupported		X		X	X	X				
User agent	X	X	X	X	X	X	X	X	X	X
Via	X	X	X	X	X	X	X	X	X	X
Warning		X	X	X	X	X				
WWW authenticate		X		X	X	X				

Accept   The *ACCEPT* header field identifies the format of the request contents. For example, in the next example, the accept header field indicates that the format for the request at the application layer is SDP, with HTML text.

```
Accept: application/sdp;level=1, application/x-private, text/html
```

When this header field is not present, the default value will always be SDP.

Accept Encoding   The meaning of this header field is almost identical to that of *ACCEPT*. This header field restricts the coding for the header field. The default value of this header field if absent is no encoding. An example of this header field is:

```
ACCEPT ENCODING:gzip
```

Accept Language   This header field is used when a language other than English is to be used for the status codes, SDP, or status responses. When this header field is absent, the default is that all languages are accepted by the server. Here is an example of the *ACCEPT-LANGUAGE* header field:

```
ACCEPT LANGUAGE: da, en-gb;q=0.8
```

Alert Info   This header field is typically used within an *INVITE* request to indicate the use of an alternate ringtone (the ringing tone used to alert the called party) to the server; however, it can also be used to indicate a different ringback tone. If this appears within the 180 RINGING response, this header field indicates the use of an alternate ringback (the tone that the calling party hears). It also provides the location of the file providing the audio ringtone or ringback tone. Here is an example of an *ALERT-INFO* header:

```
ALERT INFO: <http://www.ringtones.com/sounds/clapping.wav>
```

Allow   This is used by user agents to indicate the methods supported by the user agent. Whenever the *ALLOW* is present, all methods supported must be included or it will be interpreted that the absent method is not supported. However, the absence of an *ALLOW* does not mean that no methods are supported. For example, a query to a server to determine the server's capabilities may result in the response containing:

```
ALLOW: INVITE, ACK, OPTIONS, CANCEL, BYE
```

Authentication Info This is used as part of the authentication process and is carried within a 2*xx* response. It is used by the device to provide its authentication credentials, usually during the registration process. In an IMS environment, this is used as part of the authentication/registration process whenever a mobile is activated or changes locations. We will discuss the specifics of security and authentication/authorization later. Here is an example of how this header field is used:

```
AUTHENTICATION INFO: nextnonce="47364c23432d2e131a5fb210812c"
```

Authorization This header field contains the credentials of a user agent. These are provided typically when the device is challenged by a SIP registrar. The credentials are known to the device and the network provider only, so that fraudulent access to one's subscription is prevented. This next example shows the syntax supported for this header field:

```
AUTHORIZATION: Digest username="Travis", realm="raleigh.com",
        nonce="84a4cc6f3082121f32b42a2187831a9e",
        response="7587245234b3434cc3412213e5f113a5432"
```

Call ID The *CALL-ID* header field contains a unique identifier for *INVITEs* and *REGISTERs*. A conference call could carry multiple *CALL IDs,* one for each *INVITE*. It is the *CALL ID* that allows a server to keep track of each individual session. Any subsequent messages relating to the same session must carry the same *CALL ID.* If another session is established, a different *CALL ID* is created for that session, even if the same parties are involved.

```
CALL ID: f81d4fae-15apr-11d0-a765-00a0c91e6bf6@raleigh.com
        i:f81d4fae-15apr-11d0-a765-00a0c91e6bf6@192.0.2.4
```

Call Info This is a way for SIP to provide additional information about either the called party or the calling party. For example, the calling party may send an ICON as a representation of themselves, along with a "business card" (using vcard, for example) to the called party. The header field can be sent in either a request (in which case information about the calling party is sent) or in a response to a request (in which case information about the called party is included).

There is potential risk in using this header field. Malicious use of this header field could result in questionable content being sent to a calling or called party. The standards recommend use of this header field only if the source can be authenticated and is a trusted source.

The next example shows what this header would look like when sending both an image such as an icon and a vcard for contact information.

```
CALL INFO: <http://wwww.tekelec.com/travis/photo.jpg>;purpose=icon,
<http://www.tekelec.com/travis/>;purpose=info
```

Contact   The *CONTACT* field is used to identify the URI, or the TEL URI, or the MAIL URL that should be used to contact the user agent sending a request, associated with an established dialog. Depending on the request type, multiple URIs could be identified in the *CONTACT* field. The only stipulation is that the URI should be the same as the *TO* field. If the *TO* field contains a SIPS URI, then the *CONTACT* field should also be a SIPS URI.

When there are multiple URIs in the *CONTACT* field, the "*q*" parameter is used to indicate which order the URIs should be used. For example, the lowest numbers could be processed first, in sequential order. A redirect server uses the *CONTACT* field to indicate alternative URIs if requests are being redirected to another address.

It should also be noted that when the *CONTACT* is being used, responses could be routed directly to the address provided in the header, in essence bypassing the proxies used to originally reach the device. In cases such as IMS networks where strict routing is implemented, the *CONTACT* header simply provides additional addresses for the device; it is not used for routing purposes.

The *EXPIRES* parameter indicates in seconds when a URI provided in the header is to expire. If the value is "0," then the URI is to expire immediately. A user agent can "renew" a registration prior to expiration by sending a *REGISTER* message to the same registrar before the URI expires. A list of sessions can be renewed with one *REGISTER* message as long as the sessions all belong to the user agent, using the same *CALL ID*.

```
CONTACT: <sip:travis@ncllt2172b.raleigh.com>
```

Content Disposition   This header can be used to describe how the message body should be interpreted as an alternative to using the Session Description Protocol (SDP) in the message body. For example, the message body could carry an image to be displayed on the receiving device's display. The header would then carry the value of *RENDER,* indicating the device should simply display the image contained in the message body.

Another example for this header is when used in the response 180 RINGING. This header would use the *ALERT* parameter to indicate an alternate ringtone to be used by the receiving device to alert the user of an incoming call or message.

If this header is missing, and the *CONTENT TYPE* header carries the value "*sdp*," then the device should assume the value of *CONTENT DISPOSITION* is *SESSION* and use the SDP in the message body.

The following are possible parameter values for the *CONTENT DISPOSITION* header:

- *SESSION*   Indicates that the content is associated with a call session and therefore the message body will contain SDP.

- *RENDER*   Indicates that the message body should be rendered, or displayed to the user.

- *ICON*   Indicates the message body carries an icon that should be displayed on the receiving device.

- *ALERT*   Indicates that the message body carries a ringtone for alternative ringing.

```
CONTENT-DISPOSITION: session
```

Content Length   The *LENGTH* field indicates the total length of the message body, indicated by a decimal value indicating the number of octets. Any number is a valid value, but zero can only be used if there is no message body.

```
CONTENT-LENGTH: 265
```

Content-Transfer-Encoding   The *CONTENT-TRANSFER-ENCODING* is used when tunneling a SIP message within the message body of another SIP message. The header identifies the encoding used for the "tunneled" portion of the SIP message.

```
CONTENT-TRANSFER-ENCODING: base64
```

Content Type   The *CONTENT TYPE* header is used to identify the content media type. There must be a value in the *CONTENT TYPE* field if there is a message body. There can, however, be a *CONTENT TYPE* field and no message body (indicating an empty voice file, for example).

```
CONTENT-TYPE: application/sdp
```

CSeq   This field is used to indicate the sequence order of a message within a dialog, in the event that messages are received out of sequence. The sequence number is followed by a method (such as *INVITE*). This field is also used to differentiate requests from retransmissions.

```
CSeq: 3411 INVITE
```

Date   This header is used when devices are deriving local times and dates from the network. The device registers with the network and receives the *DATE* header in a response, which it then uses to set its own date and time. This is commonly used in many cell phones today.

```
DATE: Mon, 10 Aug 2007 14:33:03 GMT
```

Error Info   This header is used to provide additional information regarding an error response, and is used by the endpoint to provide notification to the user. For example, the error code itself could be displayed on the user's device, while the header provides the location of a recording to be played back to the user.

This provides more flexibility to the operator as to how to implement user notification and service tones/recordings.

```
SIP/2.0 404 Not Found
Error-Info: <sip:out-of-service-recording@raleigh.com
```

Event Used with the *NOTIFY* method, the *EVENT* header identifies the event that caused for a change in registration status. For example, if a Presence server has subscribed to event notification, when a subscriber device changes its registration (moves to another location resulting in a new IP address being assigned), the *NOTIFY* method sends the new registration information. The *EVENT* header then provides the reason (*reg,* for registration) the registration was changed.

```
EVENT: reg
```

Expires The *EXPIRES* header is used for multiple purposes. It allows the sender to express when an event (such as a user's registration) is to expire. For example, when the *EXPIRES* header is sent within a *REGISTER* message, the *EXPIRES* value indicates when the registration is to expire. The user must then register again prior to the *EXPIRES* value.

The use of this header varies upon its implementation. It is flexible in that it can be used to express any expiration (not just registration), depending on the method or request it is contained within. The value is expressed in decimal as seconds.

```
EXPIRES: 3
```

From The *FROM* header provides the address of the message sender. This is the logical identity, not the physical identity, and therefore must not be the IP address of a message originator. This field is used in many different ways, including rejecting or accepting calls according to user input.

The value provided in this header is for human consumption only and is not used by the network for any purpose. A device, however, may display the contents as a name display, and the user can then determine whether he or she wants to accept or reject the session.

It should be noted that this is not a secure or authenticated value. For this reason, it should not be a trusted header. The value provided in this header could easily be spoofed, leading users to accept content or other malware from a rogue source.

There is also a "tag" parameter that is added by the user agent client (UAC) when the message is created. The tag value must be a unique value to the UAC so that it can correlate this message with responses.

The *FROM* header may also contain a display name, although this is not a requirement. Likewise, the identity can be hidden, in which case the UAC will insert "Anonymous" or some other generic identity, as well as an invalid domain so as to hide the source of the message.

```
FROM: "Travis Russell" <sip:travisruss@aol.com>;tag=9hz34567sl
```

In-Reply-To This header is used for applications where calls that previously were unanswered are returned. The content of this header is the *CALL-ID* of the call being returned. It could be used by call distributors and other applications where users want control over the calls that are accepted and returned. It can also be used by voicemail systems and other like applications where the user misses a call and wants to return the missed call.

```
IN-REPLY-TO: 89222@aol.com
```

Max Forwards   *MAX-FORWARDS* is used to prevent looping of messages. When a message is created (either a request or a response), this header is set to 70. As the message passes through each proxy in the network, the value is decremented by 1, until the value reaches 0. If a proxy receives a message with a *MAX-FORWARDS* value of 0, the message is discarded.

This can also be used by systems that are measuring or monitoring the network to troubleshooting routing of SIP messages. By measuring the value of the *MAX-FORWARDS* header at various points in the network, one could derive where looping is occurring and identify proxies that are causing circular routing.

```
MAX-FORWARDS: 70
```

Min Expires   This header identifies the minimum time allowed for the expiration of a value such as *CONTACT* within a proxy, such as a registrar (the S-CSCF in the IMS). The value is a decimal number expressed in seconds.

```
MIN-EXPIRES: 60
```

MIME Version   This header identifies the version of MIME used within the message body if there is content provided in MIME format. This allows the receiving device to properly render or decode the content as received.

```
MIME-VERSION: 1.0
```

Organization   This can be used by the operator to identify the name of the operator and domain (Jack's Telephone Company for example). It can also be used by the user to identify the name of his or her company for display purposes. It identifies the name of the organization that initiated the request (or response) and is used for informational purposes (such as display on a device).

```
ORGANIZATION: Jack's Telephone Company
```

Privacy   This header is used to identify when subscribers wish to hide their identity from other networks. This could be the case when a subscriber wishes to maintain secrecy for certain communications, but of course it can be used for illegitimate purposes as well.

The header is provided in the specifications for legitimate purposes, allowing subscribers to provide a different identity in the *REQUEST-URI* than they do in the *TO* header. The *TO* header of course is displayed to the calling party, while the *REQUEST-URI* is used by the network for routing and authentication purposes.

```
PRIVACY: ID
```

Priority This header was added for providing emergency calls priority over normal traffic; however, the implementation of this header is operator specific. Therefore, since it has not been defined for global usage, it may or may not provide priority service for, say, emergency calls (such as 911 calls in the U.S.).

There is other work that is being done to define a mechanism for preemption and emergency priority services in the event of a catastrophe that are not defined here (this is still a work in progress), but this header is explained so that an operator can provide an on-net-type service if they wish.

For example, designers of an enterprise network may elect to provide an Emergency Alert System within their own private network as an additional service to their subscribers. To ensure their emergency messages were routed as a high priority, they could implement the use of the *PRIORITY* header. This would allow an enterprise network to give a higher priority to its emergency messages than it would to normal messages.

The RFC defines some suggested values for the *PRIORITY* header of *"normal," "non-urgent," "urgent,"* and *"emergency."* Their use is completely dependent on the operator, although the IETF does recommend that the use of "emergency" be limited to life-threatening situations.

The use of the *SUBJECT* header allows a text message to be included, as shown here:

```
SUBJECT: Thunderstorm Warning for Johnston County

PRIORITY: Urgent
```

Proxy Authenticate The *PROXY-AUTHENTICATE* header is provided as a challenge by a proxy (such as an application server or a registrar). When a request is sent by the device, the proxy can send a `407 Proxy Authorization Required` response, containing this header. The device then must return a response containing the *PROXY-AUTHORIZATION* header, with the proper credentials.

This could be used in cases where the operator wishes to authenticate a user every time the user accesses its call control server, to prevent unauthorized access to a server used to define call treatment and voicemail services by the subscriber.

```
PROXY-AUTHENTICATE: Digest Realm="tekelec.com,"
domain= "sip:Verizon.com", qop="auth",
nonce= "a73dd646abc898763fbade98129e763547",
opaque= "", stale= FALSE, algorithm= MD5
```

Proxy Authorization The *PROXY-AUTHORIZATION* header is used when a device is authenticating itself with the challenging proxy in response to a `407 Proxy Authentication Required`. It is within this header that the user provides his or her credentials, known only to the device and the network provider.

None of the data provided for authentication is human-consumable data. This is data that is created using the various algorithms and keys that are provided by the network provider at subscription time, and embedded within the device (usually within the UICC or SIM card).

This process is different than registration, where the network registrar performs the authentication. This procedure and its associated headers provide the operator with an added measure of security to prevent unauthorized access to network services, even after registration.

The use of this header is implementation specific and is not required, but it does add an additional level of security for the operator and an additional layer of protection to select services housed on application servers.

```
PROXY-AUTHORIZATION:Digest username= "Travis", realm= "Verizon.com",
Nonce= "c60f3082ee1212b402a21831ae"
Response= "43536ff4355tcc45567"
```

Proxy Require   This is similar to the *REQUIRE* header with the exception that it is used by proxies to communicate the extensions and capabilities that must be supported by the device. The proxy sends this header in response to requests to communicate what it requires of the user device when supporting a session.

It is different from the *REQUIRE* header, which is used between a client and a server (non-proxies). The client and the server could be resident on two devices, or they could be resident on application servers.

```
PROXY-REQUIRE:foo
```

Record Route   The *RECORD-ROUTE* header is used for strict routing within a SIP network. As a request is routed through the network, each proxy inserts this header along with its address into the request. When the request is received by the destination, it uses the *RECORD-ROUTE* to determine the route for a response.

Think of the *RECORD-ROUTE* as the headers used to create a routing list for a subscriber device. One use for this is to prevent highjacking of sessions when routing is enforced. When a device registers with the network, the *RECORD-ROUTE* is used as the *REGISTER* is routed through the network, and the various entities used to route the *REGISTER* to the registrar enter their addresses prior to forwarding to the next entity.

When the registrar receives the *REGISTER,* it then uses the *RECORD-ROUTE* headers to create a route list for the user. All responses are then sent using the same route as recorded. This route is stored as part of the registration, so that all subsequent requests and responses use the same route.

This form of strict routing ensures that a man-in-the-middle attack cannot be used to hijack a subscriber's registration, for example. It ensures that all requests and responses are sent through the same path used for the registration to reach the user.

Reply To   The *REPLY-TO* header is inserted by a user device upon receipt of a request. It is used to communicate the direct address of the device for all subsequent responses and requests throughout a dialog. This in essence would then allow responses to bypass the various proxies within the network and allow routing directly to the device.

Within an IMS domain, there may be concerns about routing directly to a device, bypassing the Call Session Control Function (CSCF) within the network. In fact, this form of loose routing is not defined within the IMS. The IMS procedures call for strict routing to ensure that all requests and responses always follow the same path used during registration.

The *REPLY-TO* header is still supported within the IMS, but its use is not necessarily for routing of responses. It is simply used to identify the direct address of the device, but requests and responses are still routed through the CSCF entities within the network. Other proxies may be bypassed, however.

```
REPLY-TO: Travis Russell <sip:travisruss@aol.com>
```

Require   This header is used by entities to identify any SIP extensions that must be supported by the other entities. This is similar to the *PROXY-REQUIRE* header discussed previously, with the exception that this is sent by devices rather than proxies.

This header is needed because SIP extensions may not be supported in every network (and consequently by every device). If a device is sending a request using specific services that are enabled through a SIP extension, then those extensions must be identified to the other endpoint to ensure the session can be supported.

```
REQUIRE: 100 REL
```

Retry-After   This header is used with the `500 Server Internal Error` and `503 Service Unavailable` responses to indicate to the requestor the duration that the server is expected to be unavailable. It can also be used with 4*xx* responses to indicate when the called party expects to be available again.

The parameters would indicate the duration of time and possibly the time when the calling party could try to reach the called party again. Optionally a message such as "In a Meeting" can be provided as part of the response to identify why the party is rejecting a call or why the called party is not available.

```
RETRY-AFTER: 19000; duration=3600
```

Route   This header is used along with the *RECORD-ROUTE* header when strict routing is implemented (as is the case in the IMS). When a request is being sent, the *RECORD-ROUTE* header records the addresses of each of the entities in the call path. The response then inserts these addresses in the *ROUTE* headers (there are typically multiple headers).

The headers are listed in the order of the route. In other words, the addresses are shown in the same order they are routed through. The routers then use this for routing the responses to the next hop in the network.

```
ROUTE: <sip:Raleigh@bellhead.com>
```

Server   This header allows the server (the recipient of a request) to communicate the software version being used by the server to process the request. There are many uses for this information, dependent on implementation.

Care should be taken with this header, as information regarding software versions could be a security risk. Hackers could use this information to obtain the version of Symbian operating system resident on a cell phone, for example, and then send a virus or Trojan to that device. Operators should use encryption to prevent this information from being read by other than the end devices in a session.

```
SERVER: Symbian OS 8.0
```

Subject   This is much like the subject line in e-mail. It is provided as a means of sharing additional information about the session for display to the user. For example, if used for an emergency broadcast feature, the *SUBJECT* header would contain the actual alert message (such as "Thunderstorm warnings for Johnston County"), while the *PRIORITY* header would contain the priority for the session.

The *SUBJECT* header can also contain text displayed on the end device for a call. The sender may wish to have the message "Answer the phone!" pop up when the call begins ringing on the called party's phone (subject to call servers supporting such a function). There are many other implementations for the *SUBJECT* header dependent on operator implementation.

```
SUBJECT: Anyone Home?
```

Supported   This header is used by either endpoint to communicate the SIP extensions and capabilities supported by the sender. This is different than *REQUIRED,* which is sent to communicate the SIP extensions that must be supported for a session.

The *SUPPORTED* header would be sent in response to *REQUIRED* or *PROXY-REQUIRE* communicating the extensions supported, or it would be included in a request/response unsolicited.

```
SUPPORTED: 100 REL
```

Timestamp   While the use of this header is not fully defined in RFC 3261, it does allow for the proxy or any other entity to enter a timestamp, which could then be used for determining round-trip time (RTT). This would require each proxy to enter the timestamp as the request/response was forwarded to the next hop, and the endpoint to have the ability to process this information.

```
TIMESTAMP: 50
```

To   This field identifies the destination for a SIP transaction. Usually this will take the form of a URI, although it can take many different forms, including a TEL URI. The address may not be the final destination for a SIP message, though, as user agents have the ultimate responsibility of determining how to interpret this field.

The address itself may be input through a human interface, or it may be provided through some other entity. For example, a user may select an address from his or her address book, in which case the *TO* field will be populated by the address book on the user's phone.

The address can be interpreted in a number of ways. For example, if the user inputs travis@tekelec.com, the user agent then assumes that a Domain Name Service (DNS) lookup is required in the Tekelec domain. The SIP message would then be sent to the domain for address resolution at that DNS.

If the user inputs a TEL URI, the proxy that receives the message would be responsible for interpreting the address and resolving the address accordingly. Likewise, each network that the request passes through would have the same opportunity, resolving the address along the way.

The *TO* field is also used to provide a display name for an incoming call or session. For example, if the URI is travis@tekelec.com, then the display of the receiving device would show Travis. This is not a requirement, so this field may not contain a display name.

Unsupported   Similar to the *SUPPORTED* header, this header is used to identify extensions that are not supported by an entity.

```
UNSUPPORTED: 100 rel
```

User Agent   This header is like the *SERVER* header providing the software version of the SIP user agent processing a request. The two headers can be used together to identify the software versions on both devices. This information can then be used by operators for profiling and other applications.

For example, in today's networks cell phones send an International Mobile Equipment Identifier (IMEI) identifying the make and model of the device, and configuration information about the device. Operators that collect this information can then use the data to track the activities of the subscriber based on that subscriber's phone make and model. They can use these statistics to better understand the behaviors of the subscribers based on their phone models.

They can also determine how the various phones are being used, and whether or not subscribers are using all capabilities of the device. If a device identified supports video, the operator can track the actual usage of all video-enabled devices to determine how many actually use this capability, when they use the capability, and so on.

This data could even be used for promotional campaigns to alert all users of certain models of devices to new features and applications that are being offered for their devices. Application servers can also process this information and use it for various applications.

```
USER-AGENT: Vista Beta1.5
```

Via   The *VIA* header is a means of recording the path that a request takes to reach its destination so that all responses follow the same path. When using loose routing, the *VIA* header ensures that responses are received by stateful proxies in the call path. In strict routing, this is used in conjunction with the *RECORD-ROUTE* header.

The difference is that the *VIA* header is used by the proxies to determine the next hop in the network for a response, while the *RECORD-ROUTE* is actually used to create a route list that will be used for routing all requests and responses to a device throughout the life of its registration. The *VIA* header is never stored as part of the registration and is only used by proxies for routing.

```
VIA: SIP/2.0/UDP pchome101@aol.com:5060; branch= z9hG4bK713a2
```

Warning Warnings indicate problems in processing the session description itself. They are different than error responses, since the session itself is being processed. The *WARNING* header contains a text description identifying the purpose of the header as defined here:

300 Incompatible network protocol

301 Incompatible network address formats

302 Incompatible transport protocol

303 Incompatible bandwidth units

304 Media type not available

305 Incompatible media format

306 Attribute not understood

307 Session description parameter not understood

330 Multicast not available

331 Unicast not available

370 Insufficient bandwidth

399 Miscellaneous warning

The definition of these warnings is outside the scope of this book. Their use is defined in RFC 3261. The preceding text values are suggested and not mandatory. Operators can define their own text values for these warnings for their own implementations.

```
WARNING: 307 tekelec.com "Session parameter 'foo' not understood"
```

WWW Authenticate This header is used as a challenge to an entity sending a request. The challenge is carried in the response to a request with this header. The response will contain the *AUTHENTICATE* header containing the proper credentials.

```
Authorization: Digest realm="raleigh.com",
domain= "sip:tekelec.com", qop="auth",
nonce="dcd98b7102dd2f0e8b11d0f600bfb0c093",
opaque="5ccc069c403ebaf9f0171e9517f40e41"
```

Table 2-2 describes how the various header fields are used, with the exception of proxies. Proxy operations differ, depending on the header field. Table 2-2 provides an overview of proxy operations by header field.

Table 2.2   Proxy Operations by Header Field

Header Field	Proxy Operations			
	A	M	D	R
Alert info	X			X
Allow				X
Call ID				X
Call info	X		X	X
Contact				
Content length	X			X
CSeq				X
Date	X			
Error info	X			
From	X	X		X
Max forwards	X			X
Organization				X
Priority	X			X
Proxy authenticate	X			X
Proxy authorization			X	X
Proxy require	X			X
Record route	X			X
Record route ($2xx$, $18x$)		X		X
Require	X			X
Route	X		X	X
To				X
Via	X	X		X
Via (copied)			X	X
WWW authenticate	X			X

Table 2-2 key:
A = The proxy can add or concatenate the header field if it is not present.
M = The proxy can modify the header field.
D = The proxy can delete the header field.
R = The proxy must be able to read the header field (cannot be encrypted).

SIP Identities

Resources within the SIP network are identified using either a SIP URI or a SIPS URI. There is no difference between the format of these two URIs. The SIPS URI indicates that the session between the user agent and the resource addressed in the URI is secure using encryption.

The SIP RFC 3261 defines the format for SIP and SIPS URIs in the form of:

```
sip:user:password@host:port;uri-parameters?headers
```

The user field identifies a resource. A resource can be a person, or it can be a server resource. The RFC does not recommend using the password field in the URI for obvious reasons. This would require publishing the user's password in plain text, naturally raising significant security concerns. Nonetheless, it has been defined but is not implemented.

There are two principles behind the use of a URI. First, one can use an IP address; however, this address should be static rather than dynamic, as the address could change when dynamic allocation is used. If a domain name is used, the receiving user agent can query the DNS to determine the IP address (common in e-mail, as an example). The URI usually takes the form of:

```
russell@tekelec.com
```

The host can be identified by using either the registered domain or the IP address. Both IPv4 and IPv6 addressing is supported in most networks today, so the format of the IP address will conform to either one of those versions. The port can also be identified but is not required. Some other examples of a SIP URI could look like:

```
russell@192.43.1.0:5060
```

There are events where the user information would not be included in the URI. For example, if the resource being addressed is a server or a router rather than a person, then the URI would not have the user information, nor would it use the "@" sign. For example:

```
sip:p4.tekelec.com
```

The URI can also take the form of a telephone number, referred to as the TEL URI. When a telephone number is used, the user information is replaced with the telephone number digits rather than the user identity. The number can be any global number using E.164 format, with a "+" preceding the number. If a local number is used, then the "+" is dropped. Hyphens can also be added to separate groups of numbers to match local notation for telephone numbers.

```
sip:+19194602172@tekelec.com
```

TEL URI supports VoIP applications where the voice is being routed through an IP network, but will need to be terminated back into the PSTN. Since the PSTN only understands telephone numbers, this format has to be supported. In a pure VoIP implementation, the use of TEL URIs is not necessary, since the SIP network is capable of routing using a standard URI.

When a call is being routed into the VoIP network from the PSTN, a TEL URI is all that can be used, since the originator in the PSTN is able to enter only telephone numbers. If the destination is an IP address, the TEL URI will have to be resolved to an IP address.

This is accomplished through the ENUM function. The purpose of the ENUM function, as you learned in Chapter 1, is to convert from telephone numbers (or TEL URI) to domain names. Once they are converted to domain names, they can be resolved by the DNS into an IP address. There are some who will implement the ENUM function as an extension of their DNS, while others will implement ENUM as a stand-alone function. Either implementation will work.

A single subscriber may possess multiple identities. Think of your own communications. You have at least one identity for e-mail. Many people have multiple e-mail addresses; one for business use and one for personal use. You also have a home phone number, a work phone number, a cell number, and possibly more.

This allows you as a subscriber to have multiple destinations for mail and voice calls, depending on the address being used. For the service provider, it allows flexibility in defining services for a single subscriber, and the ability to offer multiple services under one subscription. For each subscriber there are at least two identities known by the service provider—their private and their public identities.

Private User Identity

The *private* identity is what uniquely identifies the subscription. Remember, however, that this only identifies a subscription and cannot necessarily guarantee authenticity of the actual person using the subscription. The purpose of the private identity is to allow the operator to identify one subscription for all communications, and all services for the purposes of registration, authorization, administration, and billing.

The private identity, therefore, is not advertised to the subscriber, nor is it visible to other networks. The private identity is known only to the service provider. It may take many different forms, but it must be unique within the domain and it must be limited to one identity for each subscription. It must always be used during registration.

RFC 2486 specifies that the private identity take the form of Network Access Identifier (NAI), with the IMSI embedded as part of the address. The IMSI is already used by GSM operators as the private user identity for GSM subscribers.

The NAI as defined by the RFC is really what we have come to know as our e-mail address, where the "@" separates the username portion from the domain portion. It is the domain portion that uniquely identifies the network (and therefore the service provider).

This is why it is so important that domain names be administered by a third-party administration. If someone wishes to establish a domain name for his or her own use, that person must register with the Internet Assigned Numbers Authority (IANA) to ensure the domain is not already in use by someone else. This ensures uniqueness worldwide.

This is probably a good time to emphasize, however, that private user identities are never used for routing purposes. They are used only for administration, and so on. There are times, however, when the private user identity may need to cross network boundaries. For example, when placing a long-distance call from your home service provider, through the long-distance carrier, and back to the local operator, the local operator may be the same as your home service provider.

In this case it would be important to be able to pass along your private user identity for proper billing of the call. However, the identity must be protected from discovery by intermediate networks, so tunneling and other methods may be used.

Since the private user identity is not known by subscribers themselves, there must be a means of embedding the identity into a subscriber's device. This is commonly done today in GSM networks using the Universal Integrated Circuit Card (UICC). This is the little card inserted into the GSM phone when you obtain service from a GSM service provider. We commonly refer to this little card as the Subscriber Identity Module (SIM); however, the SIM is actually the name of the application that resides on the UICC.

The SIM provides the private user identity for the GSM service provider, and as SIP is implemented in these networks going forward, the device will continue to provide this identity in the SIP domain. This also applies to fixed-line services, where the subscriber will purchase a device and insert the SIM into the phone when that subscriber obtains services from the service provider.

This model allows subscribers to use any device they want, while uniquely identifying them to the service provider. The service provider is then able to associate the user with an authorized subscription, even though that user may be using a device he or she purchased from someplace else.

There is other information stored on the SIM application in addition to the private user identity, as we will discuss. In a perfect world, subscribers purchase a SIM application contained on a UICC from their local service provider of choice, and then use this in the various devices they purchase from the Internet, from their local electronics store, or from other service providers. The revenue stream comes from the purchase of a subscription rather than a device locked to work on the service provider network only.

Public User Identity

The *public* user identity is what subscribers use to advertise their existence. We all have one today, although ten years ago only a select group carried one. On the other hand, one could argue that we have all had a public user identity in a different form for decades: our telephone number.

The public user identity uses the same NAI form described for the private user identity, with the exception of the content, which would not consist of the IMSI (in the wireless case). The public user identity is not limited per subscriber as the private user identity is. A subscriber is likely to have multiple public user identities associated with one subscription, allowing that subscriber to use different devices for different purposes, using the same service provider (and on the same bill).

This means that every subscriber has the potential of multiple public identities to then use for personal use, business use, maybe a special hobby, or anything else he or she chooses. If you use AOL today, you will find a similar concept, where you have a primary screen name (your public user identity) but on the same account you can have multiple screen names, all billed to the same account.

Each of the public user identities will have a profile identifying the various preferences defined by the subscriber, and identifying what services are associated with the identity.

For example, I may choose to have my Blackberry service associated with my business identity, but a separate cell phone service with text and instant messaging for personal use, associated with my personal identity.

This allows the service provider to offer a lot of flexibility for its subscribers, and eliminates the need for multiple accounts for each identity. This also allows for flexible routing to the subscribers' various devices. For example, I can have calls to my business number routed directly to my Blackberry, and if there is no answer (or my Blackberry is out of range), I can have the call routed to my voicemail.

On the other hand, if my spouse calls my business number and I do not answer, I can request special routing for her call (based on her identity) so that she is routed to my personal phone rather than voicemail. The subscriber then has complete control over call treatment for each one of his or her identities.

The public user identity is used for routing, which is different than the private user identity as we discussed previously. However, the public user identity is not used for authentication. Only the private user identity can be used for authentication. Authentication is an optional function in SIP networks, unless the network is following the 3GPP IMS implementation standards. In the IMS, authentication is required anytime a subscriber accesses the network and registers his or her location.

When a subscriber activates a device and accesses the network, that subscriber's private user identity is sent along with his or her public user identity of choice (usually assigned to the device or through a login GUI on their PC) in a *REGISTER* message. This is how subscribers notify the network of their location. The *REGISTER* message is not the same as authentication. Registering does nothing more than notify the network of the location for the specified identity. Authentication requires the exchange of credentials between the device and the network. We will talk about authentication in Chapter 4.

Session Description Protocol (SDP)

The Multiparty Multimedia Session Control (MMUSIC) working group of the Internet Engineering Task Force (IETF) defined a protocol to be used in SIP networks for the announcement, description, and control of a conference utilizing multiple media types and multiple parties. A good example of such a session is a Webinar, where many parties are invited to participate in a conference utilizing text documents, PowerPoint presentations, video, audio, and other forms of media.

For this to work, there must be some form of communications to all parties regarding the type of session, the media types that must be supported to participate, and other logistical information that may be of interest to the participant. Certainly when you examine the contents of the SDP, you will see these elements are provided as a main function of the protocol.

The SDP was not originally intended for use with just SIP. In fact, SDP can be used with most any protocol including the Realtime Transport Protocol (RTP), the Hypertext Transport Protocol (HTTP), and the Simple Mail Transport Protocol (SMTP). It has been adopted for use with SIP to describe a SIP session within a SIP network, and it is defined throughout the 3GPP specifications for an IMS framework as the session description protocol for all SIP sessions.

The strength of SDP is its ability to describe a wide range of session media types. This is perhaps the reason the 3GPP has defined its use for IMS. The IMS certainly supports all media types using the SIP protocol for all session control.

The Session Description Protocol (SDP) is carried within the SIP message body itself. The SDP is what describes the session being set up by SIP. Note that not all SIP messages will contain SDP. For example, SIP messages are used to carry the content of instant messaging or Short Message Service (SMS) messages. The content of these are found in the message body of SIP, rather than in a separate packet sent through another stream (as is the case for voice and video).

The message body therefore can contain SDP describing a session found on another stream, or it can carry actual content, in which case SDP is not needed. SDP is only needed when setting up a session that will contain voice or video, or some other form of real-time content to be found on a different path.

This is analogous with Signaling System #7 (SS7), which is used in legacy networks today to set up voice calls on trunks. The actual signaling messages are carried over the same facilities separated by channels, and the SS7 message provides the details of the session being established on the voice or video channels.

The SDP protocol is actually quite simple. The headers consist of a single letter, followed by a descriptor. The descriptor is what identifies the specifications for the actual session itself. There are several descriptions provided by the SDP identifying the name of the session, the purpose of the session, the time the session starts, what media are used for the session, and any addresses and ports to be used to receive the media.

In the case of multimedia calls, SDP can describe all of the media sessions, even though they are all treated separately. For example in a Webinar session, there will be audio, voice, a whiteboard session, possibly a presentation application (such as PowerPoint), a notes application (plain text communications), a console function (identifying all of the participants and granting control over those participants to the host), and maybe even more.

Each of these media types can be described in one SIP message using the SDP to describe the various media sessions individually. Since each media type requires different parameters, the protocol must treat each one individually. In this case, the SDP describes the overall session using the session-level description, but it then uses the media descriptions to describe each individual media session.

This means that within the SDP, there might be one session description, followed by multiple media descriptions. Each of the media descriptions describes a portion of the overall session.

SDP can also contain information about the host of the session, such as contact information and bandwidth requirements for receiving the session. SDP is therefore broken into three main descriptors:

- Session-level descriptions
- Time descriptions
- Media descriptions

Session Descriptions

The *session-level* descriptions provide details about the session itself, such as the host of the session and the address for the session. There could be more than one session description within a single SIP message. In other words, one SIP *INVITE* could carry within itself SDP descriptors for more than one session. This is so when supporting a conference call, as an example; where there is need for multiple media types, each media type can be described separately.

The session descriptors in this case will always be followed by their respective time descriptions and media descriptions. Following are the "headers" for the session-level descriptions in SDP:

- v = **protocol version**   Identifies the version of SDP being supported for this session. This is used so that the receiving entity knows how to interpret the other attribute lines within the SDP.

- o = **owner/creator and session identifier**   Identifies who is initiating the session as well as identifying the session itself. Its use may not seem logical for a simple voice session between two parties, but it makes more sense when the session is a conference call or Webex session.

The format for this field is shown here:

```
o=<username><session id><version><network type><address type><address>
```

The username is all one word (cannot contain spaces) and is the user ID of the session host. For example, the username may be travis.russell. This is used along with the session ID and the rest of the parameters in this field to form a globally unique identifier for this session.

It is really up to the creator of the session how to use the <version> parameter. It was placed here to allow proxies to be able to use different versions of an announcement and determine which session announcement is the most recent. A simple method is to use a timestamp as the version, allowing the proxies to always be able to determine which is the most current.

For <network type> "IN" is used to indicate the Internet, although this certainly changes when used within an IMS environment. Most likely an IMS environment will use "IMS." Meanwhile, <address type> denotes the version of IP being used; either IPv4 or IPv6.

- s = **session name**   Again, this makes more sense when used in association with a multimedia conference bridge or Webex session. The name of the session, for example, could be something like "Introduction to SIP by Travis Russell." This might look something like:

```
s="Introduction to SIP By Travis Russell"
```

- i = **session information**   This descriptor provides additional information about the session. This is used in conjunction with the session name, and like the session name it is provided for the use by participants in the session.

There can only be one "i=" field at the session level, and only one "i=" field at the media description level. The description is a text string. It could look something like:

```
i=session on IMS
```

- $u = $ **URI of description** Identifies the URI where participants can go to receive more information about the session. Again, this makes more sense when associated with a Webex or some other similar multimedia session with multiple participants. This is an optional field, but when present it should always be found before the first media description. This field might look like:

```
u=www.tekelec.com.
```

- $e = $ **e-mail address** This allows the session owner to provide an e-mail address so that participants can contact him or her regarding the session. More than one can be provided for each session.

- $p = $ **phone number** Like the e-mail address, this is used to provide a phone number that participants can call to inquire about the session. This and the e-mail address make sense for multimedia conferences and Webex sessions but probably will not be used for simple voice communications.

- $c = $ **connection information** The connection information consists of the following:

```
c=<network type><address type><connection address>
```

The network type is the same definition as previously described. Currently "IN" has been defined for use when describing the Internet. The <address type> and <connection address> identify the actual IP address to be used for the connection, with the address type identifying it either as an IPv4 or IPv6 address.

- $b = $ **bandwidth information** Identifies the amount of bandwidth to be used for the session. It is broken into two parameters as shown here:

```
b=<modifier><bandwidth-value>
```

The modifier can be one of two values: either "CT" for conference total or "AS" for application specific. CT denotes the total bandwidth across all sites, encompassing all media used for the session. AS denotes the amount of bandwidth at a single site, from the perspective of a single application receiving the session. The bandwidth value is then expressed in kilobytes per second.

- $k = $ **encryption keys** When encryption is provided, this field is used to identify the encryption keys needed to read the payload. The standards do not identify the exact mechanism for exchanging keys, but rather a vehicle whereby encryption keys could be exchanged. The format and parameters for this field are

```
k=<method><key parameter>
```

The method may use one of several values:

- Clear = is provided without modification or alteration in this field.

- Base64 = is included in this field but has been encoded to base64 because SDP does not support its characters.

- Uri = indicates that the included URI must be used to obtain the encryption keys through additional authentication.

- Prompt = the encryption key is not included and the host should be prompted to provide the necessary encryption keys when accessing the session.

- *a* = **zero or more attribute lines**   Attributes are used as a vehicle to extend SDP for other applications. For example, if a specific application requires an attribute not already defined by the IETF, it can be called out in this parameter. Any receiving entity that does not understand the attribute simply ignores it.

  There are basically two types of attributes. An example of a property attribute might be "rcvonly." Value attributes might look something like "a=orient:landscape." These are interpreted by the receiving application, or if not understood are ignored.

Time Descriptions

Time descriptions provide additional information regarding the time of a session, including when it might be repeated. Time is always expressed in GMT to prevent confusion when crossing time zones.

- *z* = **time zone adjustments**   This allows participants to make adjustments to the time based on time zones. If the session is scheduled for eastern time, this field will identify the time zone so that the application receiving the invitation to the session can properly communicate the session time.

- *t* = **time start and stop**   This field identifies the start (the first subfield) and stop (the second subfield) times for the session. When there is only a start time, or if the end time is 0, there are special considerations that must be taken into account. If the end time is 0, then the session is considered as not "bounded." In other words, the session ends whenever terminated by the originator. The time values provided are given in Network Time Protocol (NTP) format.

- *r* = **repeat time**   This field is used to indicate when the session will be repeated again. It is represented as an offset from the start time (in other words, how long after the start time will it be repeated).

The values for the repeat time field are:

```
r=<repeat interval><active duration><list of offsets from start time>
```

This means that the repeat interval identifies the interval at which the session is to be repeated (e.g., every day), the active duration identifies the duration of the total session at each occurrence, and the offsets indicate the first and the last deltas in seconds from the start time for each of the repeated sessions. The corresponding $t=$ field would then identify the time that each of the repeated sessions are to begin. While the standards allow for the times to be represented in seconds, they also allow for short notation as shown:

```
r=7d 1h 0 25h
```

In this example, the session is repeated every 7 days (1 week), the duration is 1 hour, and the first offset is the actual first session (0), while the second offset is 25 hours.

Media Descriptions

Media descriptions provide information about the specific media for a session. When there are multimedia sessions being established, it is not uncommon for there to be one session description, one time description, and multiple media descriptions (one for each media type being supported in the session).

The format for the media description is as follows:

```
m=<media><port><transport><media formats>
```

The first field is the *media* field describing the media type. The following media types are supported for use:

- Audio
- Video
- Application
- Data
- Control

The media type could be appended as new media types are created. There is a difference between data, application, and control. Data is actual data streams sent by a particular application for processing by an end application, while application is used by whiteboard and other similar multimedia applications. Control provides for an additional control panel for an end application. Keep in mind that all applications are those that support the multimedia session, rather than any or all desktop applications.

The *port* field identifies the port to receive the session on the receiving end. The actual value of the port field will depend on the type of media being supported and the transport protocol being used to support the media.

Transport identifies the transport protocol to be used, corresponding to the $c=$ field. There are currently two values supported, *RTP/AVP,* for Real-Time Protocol using the Audio Video Profile, or *UDP.*

The *media formats* identify the formats to be used for each of the defined media types. The formats might be, for example, the type of encoding being used for voice (μ-law-encoded voice). These are identified by a number corresponding to a profile identifying the format. The numbers are then used to index to an attribute providing more detail for the format. For example, if the session is to support audio at 16 kHz, then the entire descriptor would look like:

```
m=audio 49230 RTP/AVP 99
a=rtpmap:99 L16/8000
```

The preceding example indicates the session will support audio on port 49230, using the RTP/AVP transport. The attribute line provides additional details regarding the format of the audio; in this case, it is 16 kHz audio. Attributes are described in more detail in the section that follows.

Attributes

As mentioned, attributes serve as an extension of the SDP. In the case of media descriptions, there are a number of attributes that are defined in RFC 2327, but keep in mind that various vendors may include other proprietary attributes for their own systems' use. This is perfectly acceptable and allows for more flexibility in the various entities that use SDP. The following attributes have been defined in RFC 2327:

```
a=rtpmap<payload type><encoding  name>/<clock rate><encoding parameters>
```

The payload type of this attribute indicates whether it is audio or video. In the case of audio, there may also be encoding parameters identifying the number of voice channels, although this is optional. The encoding parameters field is not used if the payload type is video.

```
a=cat:<category>
```

This attribute has been suggested for use at the session description level. An application could use this to allow the receiver of a session to sort through various sessions by category. For example, in the case of streaming audio (digital radio broadcast) the category may be the genre of the music one wants to listen to.

```
a=keywds:<keywords>
```

Keywords work a lot like the category attribute but allow for a search utility where sessions can be searched for according to specific keywords. This is also a session-level attribute.

```
a=tool:<name and version of tool>
```

This attribute is used to identify the name and the version of the tool that was used to establish a session. This is also a session-level attribute.

```
a=ptime:<packet time>
```

This is a media description-level attribute that is used to identify in milliseconds the length of time represented in received packets for a session. This is used primarily for audio to ensure proper encoding of the audio stream.

`a=rcvonly`

This can be a session- or description-level attribute and is used to set the tools to receive only when receiving a session. For example, if this is to be a Webcast session, audio is received in one direction only (receivers can listen only; they cannot have a conversation with the host of the session).

`a=sendrecv`

This is the same as the preceding attribute, with the exception that the receiving application is set in the send and receive mode. This then enables the receiver of a session to participate in the session (bidirectional audio).

`a=orient:<whiteboard orientation>`

This attribute is typically used with whiteboard applications such as those used in Webinars. The three values supported are landscape, portrait, or seascape (this is upside-down landscape).

`a=type:<conference type>`

The conference type attribute is a session-level attribute and identifies the type of conference being established. There are several suggested values:

- Broadcast
- Meeting
- Moderated
- Test

`a=charset:<character set>`

This attribute identifies the character set that is to be used to describe the character sets used when interpreting SDP. This is necessary for international networks if a character set other than ISO 10646 is to be used.

`a=sdplang:<SDP language>`

This attribute identifies the language to be used within the SDP. The default is English.

`a=lang:<language>`

Similar to SDP language, this allows a different language to be identified for a specific media descriptor. For example, the *a=sdplang:english* attribute may be found in the session description, but for a specific media description in the same session there may be a different language called out just for that medium *(a=lang:spanish)*.

```
a=framerate:<frame rate>
```

Used for video media, this attribute identifies the maximum frame rate per second to be supported.

```
a=quality:<quality>
```

This is used to suggest a level of quality for video. There are several defined values for this field:

- **10**   The best still-image possible
- **5**   Default (this value does not have a quality description)
- **0**   The worst still image quality that can be supported and is still usable

All attributes are found in either the session description or the media description levels. We have defined those found in the RFC, but remember that there can be others. Again, if an entity does not recognize an attribute, it simply ignores that attribute.

3

SIP Status Codes

Status codes communicate the status of a request, or of a session in progress. They are communicated by either entity, the client or the server agents, as well as any network entities such as proxies. The use of status codes allows the various entities to communicate with one another when there are failures or when a request is being processed.

Status codes are just that: they are responses that communicate status. To make status codes more useful and easier to manage (as well as to understand), they have been numbered and put into classifications. There are six classifications for status codes:

- Provisional
- Successful
- Redirection
- Client failures
- Server failures
- Global failures

Each of these classifications is defined in the following sections along with their respective codes. Since these are responses, they are sent when a request has been received. For example, if a user agent client (UAC) sends a request to a user agent server (UAS), then the UAS will send a response communicating the proper status of the request.

The format therefore is different than for a request. In a request, the first line of the request identifies the method, or message type. In this example you see an *INVITE* request that includes the destination of the request (or at least the next hop in the network, depending on the type of routing being used):

```
INVITE  SIP:travis.russell@tekelec.com  SIP/2.0
```

The response to this request will look like the next example. The content of the response is dependent on the status and the method of the request. This will be explained in more detail in the sections that follow describing each of the responses.

```
SIP/2.0  200 OK
```

As you can see, the first line of a request and the first line of the response are very different. I also remind you that these are displayed in clear text. In other words, bit-oriented protocols must be decoded to understand what they are, while SIP is displayed as plain English. This means that any device capable of capturing SIP messages and displaying them will be capable of displaying all of the information in the SIP message.

This could be a security or a privacy concern for some, so keep this in mind. We will discuss more about SIP security in Chapter 7. A typical call flow in a SIP network would look something like the example given in Figure 3.1.

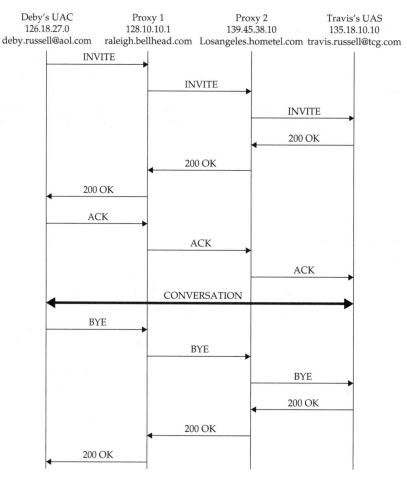

Figure 3.1   SIP simple call flow

1*xx* Provisional Codes

Provisional codes are sent to alert the client (UAC) that the client's request has been received and is being processed. Note that this may indicate that the request was received by a proxy, but not necessarily the end point (UAS).

The purpose of a provisional response is to prevent retransmission of the request. When a UAC initiates a request, it starts a timer. When this timer expires, the UAC will resend the request again, assuming that the original request was lost in the network and never received. The UAC will continue retransmission until a response is received. The UAC retransmits every 200 ms until a response is received.

The provisional response can also be used to begin a dialog between two entities. A dialog is not the same as a session. A dialog simply establishes a logical conversation (or dialog) between two entities. This dialog is then used to communicate information about a specific session. We will discuss dialogs in more detail in Chapter 5.

Since provisional responses are used by proxies to stop retransmission, the UAS (the actual destination of the request) has no knowledge of the provisional response. In fact, the UAS may send its own provisional response if it believes it will take longer than 200 ms to process the request.

When a proxy sends a provisional request, it is typically 100 TRYING, while the UAC would send 180 RINGING. The following codes have been defined in various RFCs as provisional responses:

100 Trying   The 100 TRYING response is typically sent by a proxy to prevent retransmission. This is illustrated in Figure 3.1. Note that each proxy in the path can send this provisional response, to prevent retransmission from the downstream proxy.

In other words, a proxy would send the response while processing the request (processing consists of determining the routing for the request) and then sends the request upstream to the next hop. The next (or upstream) proxy would then start its own timer, and if 200 ms is about to elapse, it will send its own provisional response. Each of the proxies within the path maintain its own retransmission timers, and each is capable of sending its own 100 TRYING response.

A typical response would look like:

```
SIP/2.0  100 TRYING
VIA: SIP/2.0/UDP  pchome101@aol.com:5060;branch=z9hG4bk74gh5
FROM: Travis Russell <sip:travis.russell@tekelec.com>
TO: Deby Russell <sip:deby.russell@aol.com>
CALL-ID: 82167534@126.18.27.0
CSEQ: 1 INVITE
CONTENT LENGTH: 0
```

180 Ringing   The 180 RINGING response is sent only by the UAS. It indicates that the UAS is attempting to alert the end user of an incoming session request. This could also be used by the UAC for triggering the transmission of ringback tones to the caller.

Ringback tones are different than alerts, because they are heard by the calling party. The 180 RINGING provisional response would then be used to cause the ringback tones to be sent by the appropriate application server.

A typical response would look like:

```
SIP/2.0  180 RINGING
VIA: SIP/2.0/UDP 128.10.10.1
FROM: Travis Russell <sip:travis.russell@tekelec.com>
TO: Deby Russell <sip:deby.russell@aol.com>
CALL-ID: 82167534@126.18.27.0
CSEQ: 1 INVITE
CONTENT LENGTH: 0
```

181 Call Is Being Forwarded   When call forwarding is implemented, this provisional response would be used to indicate the request is being forwarded to the address of record (according to the registrar or application server). The contents would indicate the address that the call was being forwarded to.

A typical response would look like:

```
SIP/2.0  181 CALL IS BEING FORWARDED
VIA: SIP/2.0/UDP 128.10.10.1
FROM: Travis Russell <sip:travis.russell@tekelec.com>
TO: Deby Russell <sip:deby.russell@aol.com>
CALL-ID: 82167534@126.18.27.0
CSEQ: 1 INVITE
CONTENT LENGTH: 0
```

182 Queued   This response is used when a party is not available but the call server is going to queue the call for later delivery. An example of this would be a service whereby the calling party is notified when the called party becomes available, and the server reconnects the session.

```
SIP/2.0  182 QUEUED
VIA: SIP/2.0/UDP 128.10.10.1
FROM: Travis Russell <sip:travis.russell@tekelec.com>
TO: Deby Russell <sip:deby.russell@aol.com>
CALL-ID: 82167534@126.18.27.0
CSEQ: 1 INVITE
CONTENT LENGTH: 0
```

183 Session Progress   This response is used to notify the originator that the request is being processed. One example could be a media gateway (MG) receiving a request, and matching the codecs required for the session with its own capabilities. It communicates this back to the media gateway controller (MGC) using this provisional response. The MGC may also send this response back to the originator to communicate that the session is progressing. There may be additional information in the headers identifying session information, depending on the sending entity.

```
SIP/2.0  183 SESSION PROGRESS
VIA: SIP/2.0/UDP 128.10.10.1
FROM: Travis Russell <sip:travis.russell@tekelec.com>
```

```
TO: Deby Russell <sip:deby.russell@aol.com>
CALL-ID: 82167534@126.18.27.0
CSEQ: 1 INVITE
CONTENT LENGTH: 0
```

2xx Successful Status Codes

There are only two codes defined for a successful request as listed here. This response is critical to a session being established (as well as a dialog) and as seen in Figure 3.1, is sent by the UAC in the backward direction to the originator.

200 OK   In order for a session to be established, the 200 OK response must be acknowledged by the UAC. The UAC upon receipt of the 200 OK will send the ACK back upstream to the UAS, and a dialog is established (as well as a session).

The 200 OK is considered a final response to a request. Once the session has been established by the receipt of the 200 OK and the transmission of the *ACK*, the session must be released by the *BYE* method.

In the case of forking proxies, a session is split to multiple destinations. This is the case in conference calls, for example. There is one *INVITE* request sent by the session originator, and a forking proxy splits the *INVITE* to multiple recipients. In this case, each of the recipients will send its own 200 OK response.

The responses are not forwarded to the session originator, however. Instead, the forking proxy will manage each of the individual responses, and send one 200 OK response back to the session originator. The forking proxy is then responsible for maintaining state of each of the recipients.

The *TAG* field in the *TO* header is used as a unique identifier in each response. The responding entity will therefore send a unique tag value in their response. This is then used by the forking proxy to maintain state of each of the recipients. The *TAG* field is one of the values used to create a dialog ID.

```
SIP/2.0  200 OK
VIA: SIP/2.0/UDP 128.10.10.1 raleighproxy.com:5060;branch=z9hG4bK63x2f
FROM: Travis Russell <sip:travis.russell@tekelec.com>tag=9hz34567sl
TO: Deby Russell <sip:deby.russell@aol.com>tag=1df789jkf
SUPPORTED: INVITE, ACK, BYE, CANCEL, OPTIONS, REGISTER
MAX FORWARDS: 70
CALL-ID: 82167534@126.18.27.0
CSEQ: 1 INVITE
CONTACT: Travis Russell <sip:travis.russell@135.18.10.10>
CONTENT TYPE: application/sdp
CONTENT LENGTH: 154
```

202 Accepted   This is used in the case of the *MESSAGE* method, which is used when sending text messages using the SIP protocol. In this case the actual text message itself is carried within the message body of the *MESSAGE* method. The response then indicates that the text message has been received by the recipient.

This response can also be sent in response to *SUBSCRIBE,* used for event notification. When an entity sends the *SUBSCRIBE* method to the registrar, the registrar then returns this response to indicate it has received the request to subscribe to event notification, and that it is processing the request. This does not mean that the event notification request has been accepted; it simply means it has been received. The *ACK* method is then used to communicate that the request has been granted.

```
SIP/2.0  202 ACCEPTED
ALLOW: ACK, INVITE, BYE, CANCEL, OPTIONS, UPDATE, MESSAGE
SUPPORTED: 100 REL
AUTHENTICATION INFO: nextnonce="47364c23432d2e131a5fb210812c"
CALL-ID: 82167534@126.18.27.0
CSEQ: 1 INVITE
CONTENT LENGTH: 0
```

3*xx* Redirection Status Codes

A redirection response is used to provide the UAC with additional addresses that the client can use to reach the subscriber. The URI for the subscriber is included as part of the response, which the client then uses to direct a request.

300 Multiple Choices   This is sent to the UAC to identify alternate locations for a subscriber. The *CONTACT* header is used to convey these additional addresses. This means that multiple *CONTACT* headers are used, one for each of the addresses.

```
SIP/2.0  300 MULTIPLE CHOICES
CONTACT: Deby Russell <sip:deby.russell@aol.com>
CONTACT: Deb <sip:druss@aol.com>
CONTACT: Deby <sip:russell@128.10.10.0>
CALL-ID: 82167534@126.18.27.0
CSEQ: 1 INVITE
CONTENT LENGTH: 0
```

301 Moved Permanently   When a subscriber moves to another address permanently, the registrar can provide this address back to the UAC. The UAC then uses this information to update any address books or directories within the device.

```
SIP/2.0  301 MOVED PERMANENTLY
CONTACT: Deby Russell <sip:deby.russell@aol.com>
CALL-ID: 82167534@126.18.27.0
CSEQ: 1 INVITE
CONTENT LENGTH: 0
```

302 Moved Temporarily   This response is used when the subscriber has moved but only temporarily. The URI is provided in the *CONTACT* header, but the UAC does not use this information to change directories or address books. The information is used only for the *INVITE* that initiated the response.

The *EXPIRES* header is also used to indicate how long the address can be stored for, in the event the UAC caches the address.

```
SIP/2.0   300 MULTIPLE CHOICES
CONTACT: Deby Russell <sip:deby.russell@aol.com>
EXPIRES: 300
CALL-ID: 82167534@126.18.27.0
CSEQ: 1 INVITE
CONTENT LENGTH: 0
```

305 Use Proxy   This response is used to indicate the address of the proxy that must be used to reach the resource being requested.

```
SIP/2.0   305 USE PROXY
CONTACT: raleigh.bellhead.com
CALL-ID: 82167534@126.18.27.0
CSEQ: 1 INVITE
CONTENT LENGTH: 0
```

380 Alternative Service   This response indicates the request was not successful, but there are alternative services that can be used. The services are defined in the message body of the response.

```
SIP/2.0   380 ALTERNATIVE SERVICE
ALLOW: INVITE, ACK, OPTIONS, CANCEL, BYE
SUPPORTED: 100 REL
ERROR-INFO: <sip:out-of-service-recording@raleigh.com>
CALL-ID: 82167534@126.18.27.0
CSEQ: 1 INVITE
CONTENT LENGTH: 0
```

4xx Client Failure Status Codes

All 4*xx* failure responses indicate problems with a session at the client (UAC) side. This does not necessarily mean that the client is at fault; it could indicate that the UAS is at fault but the UAC is reporting the issue.

As you will be able to see in the responses, the UAC uses these responses to indicate problems it is having in processing a request for a session. This response class has the largest number of defined responses out of all of the response codes.

400 Bad Request   This response indicates a syntax error contained within the message. The *REASON-PHRASE* header will provide details about the syntax error, indicating what was not acceptable.

```
SIP/2.0   400 BAD REQUEST
WARNING: 307 tekelec.com "Session parameter 'foo' not understood"
ERROR-INFO: <sip:try-again-recording@raleigh.com>
CALL-ID: 82167534@126.18.27.0
CSEQ: 1 INVITE
CONTENT LENGTH: 0
```

401 Unauthorized   This response is typically sent by the registrar to challenge a registration. When a UA sends a *REGISTER* to the registrar, the registrar can then elect to challenge the device to produce credentials as part of an authentication scheme.

The device must then send another *REGISTER* upon receipt of this response providing the proper credentials. The subsequent *REGISTER* message contains the same *CALL-ID* as the first *REGISTER,* so the registrar is able to correlate the two requests and knows that this request is in response to a challenge.

```
SIP/2.0  401 UNAUTHORIZED
ALLOW: ACK, INVITE, BYE, CANCEL, OPTIONS, UPDATE
WWW-AUTHENTICATE: authorization:Digest realm="raleigh.com",
domain= "sip:tekelec.com", qop="auth",
nonce="dcd98b7102dd2f0e8b11d0f600bfb0c093",
opaque="5ccc069c403ebaf9f0171e9517f40e41"
CALL-ID: 82167534@126.18.27.0
CSEQ: 1 INVITE
CONTENT LENGTH: 0
```

402 Payment Required   This has been reserved for future use. It could be used for example in prepaid applications where a request is being sent to a user but the user does not have adequate funds in his or her account.

```
SIP/2.0  402 PAYMENT REQUIRED
CALL-ID: 82167534@126.18.27.0
CSEQ: 1 INVITE
CONTENT LENGTH: 0
```

403 Forbidden   This response is used when a subscriber wishes to reject an incoming call, for example. Since the session has not been established (the user is being alerted, so the *ACK* has not yet been received), the *BYE* method cannot be used to release the session.

This could be used for other purposes. If a UAS rejects a session for any reason, this response can be given to reject the request. This would then effectively release the dialog prior to an *ACK* being received by the UAC.

If a user fails authentication, the registrar will send this response. For example, in IMS networks the S-CSCF (which acts as the registrar) will send this response after challenging a subscriber and then failing the second attempt to register. This could happen if a subscriber is already registered and attempts to register again.

```
SIP/2.0  403 FORBIDDEN
CALL-ID: 82167534@126.18.27.0
CSEQ: 1 INVITE
CONTENT LENGTH: 0
```

404 Not Found   This response is sent if the subscriber identified in the request is not known within the domain that the request was made. It can also mean that the domain identified in the URI is not known by the domain that the request was sent to, as would be the case if the network could not reach the specified domain.

```
SIP/2.0  404 NOT FOUND
ALLOW: ACK, INVITE, BYE, CANCEL, OPTIONS, UPDATE
RETRY-AFTER: 10
CALL-ID: 82167534@126.18.27.0
CSEQ: 1 INVITE
CONTENT LENGTH: 0
```

405 Method Not Allowed The method sent is not allowed. For example, if the method sent was *MESSAGE* but the subscriber identified in the request-URI does not subscribe to messaging services, the network would return this response to indicate the subscriber addressed is not allowed to receive methods of this type. The response also contains the *ALLOWED* header identifying the methods that the subscriber is allowed to receive.

```
SIP/2.0  405 METHOD NOT ALLOWED
ALLOW: ACK, INVITE, BYE, CANCEL, OPTIONS, UPDATE
SUPPORTED: ACK, INVITE, BYE, CANCEL, OPTIONS, UPDATE
CALL-ID: 82167534@126.18.27.0
CSEQ: 1 INVITE
CONTENT LENGTH: 0
```

406 Not Acceptable This response indicates that the destination of the request cannot support the requirements that are specified in the *ACCEPT* header.

```
SIP/2.0  406 NOT ACCEPTABLE
ACCEPT: application/sdp;level=1, application/x-private, text/html
CALL-ID: 82167534@126.18.27.0
CSEQ: 1 INVITE
CONTENT LENGTH: 0
```

407 Proxy Authentication Required This is similar to the response 401 but requires the UAC to first go through authentication with the proxy (for example, prior to granting access to a communications channel off a telephony gateway).

```
SIP/2.0  407 PROXY AUTHENTICATION REQUIRED
ALLOW: ACK, INVITE, BYE, CANCEL, OPTIONS, UPDATE
SUPPORTED: ACK, INVITE, BYE, CANCEL, OPTIONS, UPDATE
PROXY-AUTHENTICATE: Digest Realm="tekelec.com,"
domain= "sip:Verizon.com", qop="auth",
nonce= "a73dd646abc898763fbade98129e763547",
opaque= "", stale= FALSE, algorithm= MD5
WWW-AUTHENTICATE: Digest realm="raleigh.com",
domain= "sip:tekelec.com", qop="auth",
nonce="dcd98b7102dd2f0e8b11d0f600bfb0c093",
opaque="5ccc069c403ebaf9f0171e9517f40e41"
CALL-ID: 82167534@126.18.27.0
CSEQ: 1 INVITE
CONTENT LENGTH: 0
```

408 Request Timeout   This is sent when a UAS is unable to send a response within a suitable time. This happens when the UAS could not find the subscriber in time.

```
SIP/2.0  408 REQUEST TIMEOUT
CALL-ID: 82167534@126.18.27.0
CSEQ: 1 INVITE
CONTENT LENGTH: 0
```

410 Gone   This is sent to indicate a UAS is no longer available at the server, but no forwarding address is known. The condition is permanent. This is different than response 404, which is sent if the UAS cannot determine if the condition is permanent.

```
SIP/2.0  410 GONE
CALL-ID: 82167534@126.18.27.0
CSEQ: 1 INVITE
CONTENT LENGTH: 0
```

413 Request Entity Too Large   This indicates the request body is too large for the UAS to process.

```
SIP/2.0  413 REQUEST ENTITY TOO LARGE
ALLOW: ACK, INVITE, BYE, CANCEL, OPTIONS, UPDATE
RETRY-AFTER: 10
CALL-ID: 82167534@126.18.27.0
CSEQ: 1 INVITE
CONTENT LENGTH: 0
```

414 Request URI Too Long   This response means the request-URI is longer than the server is willing to process.

```
SIP/2.0  414 REQUEST URI TOO LONG
CALL-ID: 82167534@126.18.27.0
CSEQ: 1 INVITE
CONTENT LENGTH: 0
```

415 Unsupported Media Type   This indicates the message body cannot be processed by the UAS. The UAS will return accepted formats in the *ACCEPT, ACCEPT-ENCODING,* or *ACCEPT LANGUAGE* headers.

```
SIP/2.0  415 UNSUPPORTED MEDIA TYPE
ACCEPT: application/sdp;level=1, application/x-private, text/html
ACCEPT LANGUAGE: da, en-gb;q=0.8
ERROR-INFO: <sip:out-of-service-recording@raleigh.com>
ALLOW: INVITE, ACK, OPTIONS, CANCEL, BYE
SUPPORTED: 100 REL
CALL-ID: 82167534@126.18.27.0
CSEQ: 1 INVITE
CONTENT LENGTH: 0
```

416 Unsupported URI Scheme   A proxy uses this response to reject a request when it cannot understand the URI being sent in the request. This occurs when the URI scheme being used cannot be deciphered by the UA.

```
SIP/2.0  416 UNSUPPORTED URI SCHEME
CALL-ID: 82167534@126.18.27.0
CSEQ: 1 INVITE
CONTENT LENGTH: 0
```

420 Bad Extension   One example for this response is that the UA does not understand the *OPTIONS-TAG* field within the *PROXY-REQUIRE* or *REQUIRE* header. The response will include an *UNSUPPORTED* header indicating the received extensions it does not support.

```
SIP/2.0  420 BAD EXTENSION
ALLOW: INVITE, ACK, OPTIONS, CANCEL, BYE
SUPPORTED: 100 RELR
ERROR-INFO: <sip:out-of-service-recording@raleigh.com>
CALL-ID: 82167534@126.18.27.0
UNSUPPORTED: P-CHARGING-FUNCTION-ADDRESSES
CSEQ: 1 INVITE
```

421 Extension Required   This response is sent when the UAS needs an extension that is not identified in the *SUPPORTED* header. If possible, the UAS will process the request using basic capabilities rather than reject the request.

```
SIP/2.0  421 EXTENSION REQUIRED
ALLOW: INVITE, ACK, OPTIONS, CANCEL, BYE
SUPPORTED: 100 REL
ERROR-INFO: <sip:out-of-service-recording@raleigh.com>
CALL-ID: 82167534@126.18.27.0
UNSUPPORTED: P-CHARGING-FUNCTION-ADDRESSES
CSEQ: 1 INVITE
```

423 Interval Too Brief   This is returned to the client if the expiration time identified in the *CONTACT* header is too short of an interval. The expiration time is communicated in the *CONTACT* header only in *REGISTER* methods.

```
SIP/2.0  423 INTERVAL TOO BRIEF
CALL-ID: 82167534@126.18.27.0
CSEQ: 1 INVITE
CONTENT LENGTH: 0
```

480 Temporarily Unavailable   This is used to indicate that the session can be connected, but the subscriber is not available. There are numerous reasons why the subscriber would be unavailable, including invocation of Do Not Disturb feature.

There is also an option to provide more details in a user-definable reason phrase. This is implementation specific. There is also an option to place a value in the *RETRY-AFTER* header to indicate when the session can be attempted again. This is also implementation dependent, and both the *REASON-PHRASE* and the *RETRY-AFTER* would be populated by an application on the subscriber device.

```
SIP/2.0  480 TEMPORARILY UNAVAILABLE
ALLOW: ACK, INVITE, BYE, CANCEL, OPTIONS, UPDATE
RETRY-AFTER: 10
CALL-ID: 82167534@126.18.27.0
CSEQ: 1 INVITE
CONTENT LENGTH: 0
```

481 Call/Transaction Does Not Exist   This indicates the session identified does not exist. The client therefore has terminated the session. The UAS sends this in the backward direction in response to a request regarding an existing session. This is also sent when a *BYE* method is received, and the UA cannot correlate it to any session ID or dialogs.

One example could be where a proxy receives a request for a session, when the session has already been canceled using the *CANCEL* method. The *CANCEL* is sent by the UAC when it wishes to release a session prior to receiving an *ACK*. Since the *ACK* has not been received yet, a dialog has not been established between the two entities, and the *BYE* method cannot be used to release the session. This response then indicates that the *CANCEL* method has already released the session and the UAS is unable to identify the session being requested.

An alternative use for this response could be 481 SUBSCRIPTION DOES NOT EXIST, used with the *SUBSCRIBE* and *NOTIFY* methods to indicate that the subscription that has been identified in either one of these methods has not yet been defined and granted, and therefore it does not exist.

For example, a new subscription may have been entered at a Point of Sale (POS), but the subscription has not been provisioned throughout the network, resulting in some entities having knowledge of the subscription and requesting event notification.

```
SIP/2.0  481 CALL/TRANSACTION DOES NOT EXIST
CALL-ID: 82167534@126.18.27.0
CSEQ: 1 INVITE
CONTENT LENGTH: 0
```

482 Loop Detected   The server detected a loop by finding a SIP message with its own address in the *VIA* header. If it determines that looping may have occurred, the proxy can also calculate the *BRANCH ID* and match it against the *BRANCH ID* in the received *VIA* header.

Note that this is different than response 483, which identifies there are too many hops for a particular message as determined by the hop counter. That condition may or may not imply circular routing, whereas this response indicates circular routing.

```
SIP/2.0  482 LOOP DETECTED
CALL-ID: 82167534@126.18.27.0
CSEQ: 1 INVITE
CONTENT LENGTH: 0
```

483 Too Many Hops   Indicates that the *MAX-FORWARDS* header contains a value of *0*, meaning that the request has gone into circular routing and cannot reach its destination.

```
SIP/2.0  483 TOO MANY HOPS
CALL-ID: 82167534@126.18.27.0
CSEQ: 1 INVITE
CONTENT LENGTH: 0
```

484 Address Incomplete   The *REQUEST-URI* was not complete.

```
SIP/2.0  484 ADDRESS INCOMPLETE
ALLOW: ACK, INVITE, BYE, CANCEL, OPTIONS, UPDATE
SUPPORTED: 100 REL
ERROR-INFO: <sip:out-of-service-recording@raleigh.com>
CALL-ID: 82167534@126.18.27.0
CSEQ: 1 INVITE
CONTENT LENGTH: 0
```

485 Ambiguous   This indicates that the *REQUEST-URI* was too ambiguous for routing. The message could also include possible alternatives in the *CONTACT* header, but there are concerns that implementation of this capability may pose security concerns.

```
SIP/2.0  485 AMBIGUOUS
SUPPORTED: REGISTER, INVITE, ACK, BYE, CANCEL, OPTIONS, UPDATE
CONTACT: raleigh.bellhead.com
CALL-ID: 82167534@126.18.27.0
CSEQ: 1 INVITE
CONTENT LENGTH: 0
```

486 Busy Here   This is returned when the called user is unable or unwilling to receive a call at the device receiving the request. This is not a global response for the user in general. Possibly, requests are being sent to multiple addresses for the same session.

For example, a subscriber may have requests being sent to multiple devices (e-mail or voicemail devices for example). If one of those devices happens to be busy, there may still be other devices for the same user that are not busy and can accept the session. The device that is busy would return this response, while the other devices would return a *2xx* response.

This response may also carry the header *RETRY-AFTER* indicating a better time to try the request again.

```
SIP/2.0  486 BUSY HERE
ALLOW: ACK, INVITE, BYE, CANCEL, OPTIONS, UPDATE
RETRY-AFTER: 10
CALL-ID: 82167534@126.18.27.0
CSEQ: 1 INVITE
CONTENT LENGTH: 0
```

487 Request Terminated   If the expire timer expires for a request before a final answer can be delivered, the UAS will deliver this response to indicate the session has been terminated and the request will have to be sent again.

```
SIP/2.0  487 REQUEST TIMEOUT
CALL-ID: 82167534@126.18.27.0
CSEQ: 1 INVITE
CONTENT LENGTH: 0
```

488 Not Acceptable Here   This response is sent when rejecting a request for a resource identified within the request. The response will also provide the reason code for rejecting the request. This is different than *606 Not Acceptable,* which addresses the entire request rather than just the request for a specified resource.

For example, a media gateway controller function (MGCF) may send this response when the required codec identified in the SDP portion of the *INVITE* is not supported by the MGCF. The MGCF may include the SDP in the response to identify the codec(s) that are supported by the MGCF.

This can also be sent when rejecting a re-invite to reject the session description. The *WARNING* header will identify the reason for rejecting the re-invite. Here is an example:

```
SIP/2.0  488 NOT ACCEPTABLE HERE
WARNING: Media Style Not Available
MAX FORWARDS: 70
CALL-ID: 82167534@126.18.27.0
CSEQ: 1 INVITE
CONTENT TYPE: application/sdp
CONTENT LENGTH: 154
```

489 Bad Event   This is used when an "event package" (or event type) defined in a *SUBSCRIBE* or *NOTIFY* message is not understood by the UAS. The UAS is therefore unable to decipher what the event is that the UAC is attempting to either subscribe to or provide notification of.

```
SIP/2.0  489 BAD EVENT
CALL-ID: 82167534@126.18.27.0
CSEQ: 1 INVITE
CONTENT LENGTH: 0
```

491 Request Pending   The UA will send this response when it receives a request while waiting for a previous request to be processed (within the same dialog). This prevents "glare" from occurring.

```
SIP/2.0  491 REQUEST PENDING
CALL-ID: 82167534@126.18.27.0
CSEQ: 1 INVITE
CONTENT LENGTH: 0
```

493 Undecipherable   The UAS sending this response has received an encrypted request that the UAS is unable to decrypt because it does not possess the proper encryption keys.

```
SIP/2.0  493 UNDECIPHERABLE
CALL-ID: 82167534@126.18.27.0
CSEQ: 1 INVITE
CONTENT LENGTH: 0
```

5*xx* Server Failure Status Codes

Server failures indicate problems at the UAS when attempting to process requests and responses. The UAS or the UAC can originate these responses, depending on the circumstances. For example, if a UA receives a second *INVITE* within a single dialog, and the second *INVITE* has a lower *CSeq* number than the first *INVITE*, and the UA has not sent a response to the first *INVITE*, than the UA sends a 500 SERVER INTERNAL RESPONSE response.

All 5*xx* responses are considered as final responses.

500 Server Internal Error   This response usually indicates messages have been received out of sequence. The response will typically include a RETRY-AFTER header with a value between 0–10 seconds (determined randomly by the UA).

```
SIP/2.0  500 SERVER INTERNAL ERROR
ALLOW: ACK, INVITE, BYE, CANCEL, OPTIONS, UPDATE
SUPPORTED: 100 REL
ERROR-INFO: <sip:out-of-service-recording@raleigh.com>
RETRY-AFTER: 19000; duration=3600
CALL-ID: 82167534@126.18.27.0
CSEQ: 1 INVITE
CONTENT LENGTH: 0
```

501 Not Implemented   The UAS sends this response if the method being requested is not implemented at the UAS. This is different than 405 METHOD NOT ALLOWED because the 405 response indicates the method is implemented at the UAS but does not support it.

```
SIP/2.0  501 NOT IMPLEMENTED
CALL-ID: 82167534@126.18.27.0
CSEQ: 1 INVITE
CONTENT LENGTH: 0
```

502 Bad Gateway   A downstream server sent an invalid response while the proxy was acting as a gateway.

```
SIP/2.0  502 BAD GATEWAY
CALL-ID: 82167534@126.18.27.0
CSEQ: 1 INVITE
CONTENT LENGTH: 0
```

503 Service Unavailable   This response is sent when a proxy is out of service for maintenance or is overloaded. The proxy may indicate a time to retry using the *RETRY-AFTER* header in the response, but this is not a requirement. A client receiving this response should choose an alternate proxy, but it should not send any additional requests to the proxy until after the expiration of the *RETRY-AFTER* value.

Another example could be a case where an MGCF is unable to provide the requested service because the media gateway (MG) does not appear to have the required codecs to support its request.

If this is received by a stateful proxy, the proxy does not forward the message but rather generates a 500 response and sends that upstream.

```
SIP/2.0  503 SERVICE UNAVAILABLE
ALLOW: ACK, INVITE, BYE, CANCEL, OPTIONS, UPDATE
SUPPORTED: 100 REL
ERROR-INFO: <sip:out-of-service-recording@raleigh.com>
RETRY-AFTER: 19000; duration=3600
CALL-ID: 82167534@126.18.27.0
CSEQ: 1 INVITE
CONTENT LENGTH: 0
```

504 Server Timeout This response is sent if a UAS does not receive a response in a timely manner. This is used only if waiting for a response from a UAS in another network domain. If there is an internal UAS, the 408 REQUEST TIMEOUT is used to indicate no response within the time specified in the *EXPIRES* header.

```
SIP/2.0  504 SERVER TIMEOUT
CALL-ID: 82167534@126.18.27.0
CSEQ: 1 INVITE
CONTENT LENGTH: 0
```

505 Version Not Supported The UAS does not support the SIP version identified in the request start line.

```
SIP/2.0  505 VERSION NOT SUPPORTED
CALL-ID: 82167534@126.18.27.0
CSEQ: 1 INVITE
CONTENT LENGTH: 0
```

513 Message Too Large This response is used to indicate the received message was too large for the server to process.

```
SIP/2.0  513 MESSAGE TOO LARGE
CALL-ID: 82167534@126.18.27.0
CSEQ: 1 INVITE
CONTENT LENGTH: 0
```

6*xx* Global Failure Status Codes

All other failures within the network fall under the category of global failures. These failures are circumstances that are network-wide rather than isolated to a specific entity. These are also final responses.

The focus of 6xx responses is different than those of 4xx and 5xx in that these responses focus on sessions. For example, 4xx responses focus on client failures, while 5xx responses focus on server issues, and why a specific session could not be supported by either entity.

The 6xx responses focus on a subscriber regardless of session attempts and provide information on the status of a subscriber within the network. Of course, these responses are sent in response to requests, but they do not necessarily have to be associated with session-based requests.

600 Busy Everywhere   This response is used only if the receiving UAS knows that the user is not available at any other location in the network. In an IMS network this could be a presence server. In a traditional SIP network, it would have to be an entity such as a stateful proxy.

The *RETRY-AFTER* header could be used to indicate times to retry the request. The response also contains a reason phrase; otherwise, the UAS responds with 603 DECLINE instead. If there is an alternative resource such as voicemail available for the subscriber, then the response 486 BUSY HERE is sent as an alternative.

```
SIP/2.0  600 BUSY EVERYWHERE
ALLOW: ACK, INVITE, BYE, CANCEL, OPTIONS, UPDATE
RETRY-AFTER: 19000; duration=3600
CALL-ID: 82167534@126.18.27.0
CSEQ: 1 INVITE
CONTENT LENGTH: 0
```

603 Decline   This is sent as an alternative to 600 when no reason phrase is being sent. It is used to indicate that the user is not available or does not wish to accept a session invite. The *RETRY-AFTER* header is used to indicate a time to attempt the request again. The UAC in this instance has been effectively reached, but there is no other resource to accept the request and the subscriber has rejected the request.

```
SIP/2.0  603 DECLINE
ALLOW: ACK, INVITE, BYE, CANCEL, OPTIONS, UPDATE
RETRY-AFTER: 19000; duration=3600
CALL-ID: 82167534@126.18.27.0
CSEQ: 1 INVITE
CONTENT LENGTH: 0
```

604 Does Not Exist Anywhere   This response is sent only if the server knows that the user identified in the request URI no longer has service in the domain identified. For example, a subscriber may have ported to another network or had his or her service disconnected.

```
SIP/2.0  604 DOES NOT EXIST ANYWHERE
CALL-ID: 82167534@126.18.27.0
CSEQ: 1 INVITE
CONTENT LENGTH: 0
```

606 Not Acceptable   This response means that the UAS is willing to accept an invitation but is unable to support the session as described. The *WARNING* header provides additional details as to why the UAS cannot support the request.

```
SIP/2.0  606 NOT ACCEPTABLE
CALL-ID: 82167534@126.18.27.0
CSEQ: 1 INVITE
CONTENT LENGTH: 0
```

Registration Procedures in a SIP Network

One of the biggest advantages to SIP networks is the ability to move about the network and change locations. This form of mobility requires recording the location address (IP address) someplace in the network when the user changes locations. This is so the network will know to what address it should be routing requests to at any given time.

This of course is not possible without notifying the network of the location (IP address) each and every time the address is changed. The registration process in SIP was developed for just that: the sharing of physical addresses each time the user's address changes.

Registration can be about more than just registering a new IP address. It should also be an opportunity for the network to authenticate the users accessing its network. This is something that is missing in many VoIP networks today, but something that is quite simple to accomplish within the SIP domain if it is implemented.

Of course the physical address is not the same as the public address. The public address (like our e-mail addresses) is what other people use to reach us no matter where we are. This public address must be resolved to our physical IP address, which then requires a function in the network to maintain this information.

This function is the registrar function, which can be centralized in the network or distributed throughout the network. The registrar is the function that matches our public URI to our physical IP address. The proxies in the network then access this function to obtain our physical address prior to routing requests/responses to our devices.

This is probably the most exciting feature of VoIP. For example, I can launch a VoIP application on my laptop, and no matter where I am in the world, telephone calls will always be routed to me as long as I am connected to the network (such as the Internet). This is a function that has made Skype very popular among many business travelers.

Even video can be added to make the conversation even more interesting. Conference calling using small porcameras attached to PCs is common nowadays, and a great way to reach out to friends and family in other cities. No long-distance fees! And all your friends and family need is your public identity, such as a telephone number (TEL URI).

No need to know the address where you are at any given time, which is the case with switched network services today. Geographical portability is a concept long talked about but always said to be impossible by the traditional service providers. It took the likes of Skype and Vonage to prove that not only was it feasible, but consumers would buy it.

Today this type of service is becoming commonplace, and many service providers are migrating their networks to IP. As they make this migration, addresses will change from traditional telephone numbers to URIs. These public address will then be associated with physical addresses in the network in real time.

The association of a physical address to a public address is referred to as a *binding*. The binding process requires a SIP message sent to the SIP registrar, which is then responsible for storing the binding (the coupling of the physical address with the public address) and for storing the location information in a location server. It is this location server that then becomes available for other network entities to learn our device address for routing.

This chapter is about more than just registration, however. It is also about the migration from switched networks to IP networks. This is a migration that will take many years to complete, so we should not expect to see this happen overnight.

In fact, service providers have already begun the migration to IP, but they are leaving their legacy networks intact for the most part. This means they are putting in IP backbones first and then slowly moving the rest of their network over.

The legacy switched network will need to interwork with the newer IP network for many years to come, which presents special challenges. Think of all the service platforms in the network today that are based on Intelligent Network (IN) technology. These platforms do not work with SIP; they only support SS7.

This means interworking between SIP and SS7 will be necessary. We will talk about this interworking a little later in this chapter. The interworking is an important factor because as we mentioned before, it may take a long time before IP replaces the entire network.

There must also be functions that act as gateways between the two domains, so that services will continue to work in both the switched network as well as the SIP network, without having to replace the platforms. These gateways work in both domains, managing the SS7 connection on one side and the SIP connection on the other, making them stateful proxies in the SIP domain. We will discuss more about these functions as well.

Basic Registration

As discussed, the user device must share its physical address with the registrar in the network. Along with this "registration" is the public identity that is to be bound to the physical address. Keep in mind that the public URI can change physical addresses many times as a subscriber moves about the network, so the binding of addresses may change frequently (and rapidly if the subscriber is driving down the interstate).

The registration process begins when a subscriber access the IP network and obtains his or her IP address from the network. How this mechanism works is outside the scope of this book, so we won't go into those details. Once the device has obtained its IP address, a SIP application is launched and the device then sends its address information to the SIP registrar.

The *REGISTER* message is routed through the network according to local policy, using loose routing typically. In simple SIP networks, the *RECORD-ROUTE* headers are not used with this method, and therefore any *RECORD-ROUTE* headers will be ignored by the proxies. However, in IMS networks strict routing is used and the *RECORD-ROUTE* headers are used for recording the path of a registration.

The UAC within the device will use the SIP *REGISTER* method for sending both the IP address as well as the public identity to be bound with the IP address. From that point forward all requests and responses addressed to that public identity will be sent to the IP address specified, until either the registration expires, or the location changes and the device registers the new address.

The binding occurs when the UAC places the public identity (or address of record) in the *TO* header and the IP address in the *CONTACT* header. This information is then stored in the location server behind the registrar. If the device is trying to bind multiple public identities to the same IP address, it can send a *REGISTER* message with multiple *CONTACT* headers containing the other public identities.

The *REGISTER* method can be used by only one device. In other words, one *REGISTER* cannot be used to register multiple devices; only multiple identities to the same device. The registration is then maintained until expiration, which is defined in the *EXPIRES* header.

The *EXPIRES* field provides the amount of time that a registration is to be kept. When more than half of the time has lapsed, the UAC will either send a new *REGISTER* to renew the registration time or allow the expiration of the current registration. Default expiration time is one hour unless specified otherwise.

The *EXPIRES* field in the *CONTACT* header applies to each address in the header. The UAC can also manually expire the registration by sending the *REGISTER* message again, containing the *CONTACT* header with the address to be expired, and an *EXPIRES* field with a value of 0.

To retrieve the current bindings from the registrar, a *REGISTER* is sent without the *CONTACT* header. The address of record is provided (public address) so that it can be correlated to the bindings currently stored. The response will then contain the *CONTACT* header for each IP address currently bound to the public identity.

The location of the registrar can be discovered in a number of ways. One method is to configure the address of the registrar within the device itself. This makes sense for some types of devices that will most likely be in a fixed location, such as an IP access device supporting VoIP. The registrar address is shipped in the configuration of the device so that when the subscriber plugs his or her device into their broadband access, the device automatically sends the registration information to the registrar.

Another method is to route the registrations according to the domain portion of the public identity. For example, my own address of travisruss@aol.com would force registration messages to be routed to aol.com. The domain is identified in the request-URI of the message.

IP Multicast is another method that can be used in some cases. The address then becomes sip:mcast.net, which then gets routed to the IPv4 address of 224.0.1.75. The proxies within the network then are responsible for routing to the correct location according to the method *REGISTER*.

A typical *REGISTER* message would look something like the next example. Notice that the *TO* and the *FROM* headers are the same. Also notice that the request-URI only contains the domain, since the message is not being routed to another subscriber or application server.

```
REGISTER  sip: tcg.com  SIP/2.0
TO: travisruss@tcg.com
FROM: travisruss@tcg.com
CONTACT: travis.russell@135.18.10.10
CALL-ID: 82163456@tcg.com
CSeq: 1001 REGISTER
```

The response that is sent in response to a *REGISTER* message is the 200 OK, containing the *CONTACT* header with the current bindings for the address being addressed at that registrar. A typical registration process is shown in Figure 4.1.

The registrar is really nothing more than a server acting as a UAS, and storing the bindings for each registered subscriber. As with any UAS processing a request, the registrar can only process one *REGISTER* at a time from any one device. The registrar also shares this information with other proxies that request addresses for subscribers.

One measure that should be implemented in all SIP networks is authentication. This requires the exchange of credentials between the device and the network. The registrar should authenticate every device prior to registering the device. Sadly, this is a practice that is not widespread in today's VoIP networks. This is one of the reasons we see many different cases of VoIP fraud from unauthorized access to the network.

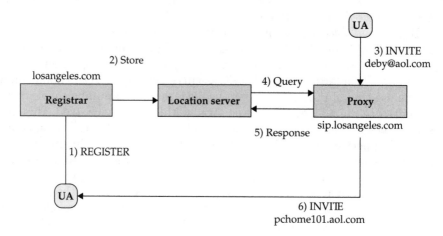

Figure 4.1 Typical registration process

Authentication consists of a simple challenge when a device is registering with the network. Instead of accepting the *REGISTER* message when it is sent, the registrar rejects the request using the 401 UNAUTHORIZED response.

The user device will then resend the *REGISTER* message containing the proper credentials (authentication keys), which have been embedded within the device. We talk about this in a number of places, but the detailed description of this process can be found in Chapter 7.

Event Notification

Within a SIP network there are other functions that may need to know when a subscriber's registration changes, so that the application can continue to reach the subscriber. One such service is presence. A presence application provides those subscribed near real-time status of the availability of a subscriber. Think of your buddy list as one example of presence.

A presence application must be notified when a subscriber's location or status changes. SIP uses event notification to provide these applications updates when a subscriber changes his or her registration. The application server providing the service must establish a dialog with the registrar for event notification to work properly.

The application server creates the dialog by using the *SUBSCRIBE* method. The *SUBSCRIBE* method identifies the subscriber that the application server wants notification on, and subscribes the application server to the event notification function.

When the state of a subscriber changes, the registrar will notify any application servers (or other user agents) of the new status. The registrar uses the *NOTIFY* method for providing state change information. The actual procedures differ according to the type of event package that the application's server has subscribed to.

Message Waiting Indication (MWI)

One good example of event notification is for message waiting indication. Message waiting indication is a common telephony feature used with voicemail platforms. The voicemail platform sends notification to the subscriber phone to indicate the subscriber has a message waiting. The notification method can be a light on the phone, a display message, or even a stutter dial tone.

In a SIP network there is no method for duplicating this function, so event notification is used. When a subscriber device registers in the network, it must then send the *SUBSCRIBE* method to the message platform. The subscription to event notification is only good for a specified duration (usually a few hours or a few days).

This means that the device must continue to send the *SUBSCRIBE* prior to the event notification subscription expiring. The duration of the subscription is defined in the *EXPIRES* header the same as in the *REGISTER* message. It is then up to the subscriber device to ensure that it sends another *SUBSCRIBE* well in advance of the expiration. Usually this is when half of the expiration time has been met.

Next is an example of message sequence between a subscriber device and a voice-mail platform (as an example).

```
SUBSCRIBE sip:travis.russell@tcg.com SIP/2.0
TO: <sip:travis.russell@tcg.com>
FROM: <sip:travis.russell@tcg.com>;tag=31987
DATE: Fri, 16 Nov 2007 02:45:04 GMT
CALL-ID: 763472@travis-cell. com
CSeq: 1 SUBSCRIBE
CONTACT: <sip:travis.russell@tcg.com>
EVENT: message-summary
EXPIRES: 86400
ACCEPT: application/simple-message-summary
CONTENT-LENGTH: 0
```

In this example, the *EVENT* header identifies the event notification package this is applicable to, and in the *ACCEPT* header the application is identified. Notice also the *EXPIRES* header. This is a simple example and may or may not be different in actual practice, but the intent is to provide a basic example for reference. This message is followed by a 200 OK response, but no *ACK* is required.

Whenever the voice mailbox of the subscriber changes state, the voicemail system sends the *NOTIFY* message to notify the subscriber of a message waiting. There are a number of ways the device then acts upon this message, but it is dependent on the device to determine how to notify the subscriber.

The *NOTIFY* provides notification in the form of plain text. The device must be able to read this plain text then but does not necessarily display the contents of the *NOTIFY*. Here is an example of a typical *NOTIFY* message:

```
NOTIFY sip:travis.russell@tcg.com SIP/2.0
TO: <sip:travis.russell@tcg.com>
FROM: <sip:travis.russell@tcg.com>;tag=31987
DATE: Fri, 16 Nov 2007 02:45:04 GMT
CALL-ID: 763472@travis-cell. com
CSeq: 20 NOTIFY
CONTACT: <sip:travis.russell@tcg.com>
EVENT: message-summary
SUBSCRIPTION-STATE: active
CONTENT-TYPE: application/simple-message-summary
CONTENT-LENGTH: 99
MESSAGES-WAITING: yes
MESSAGE-ACCOUNT: sip:travis.russell@tcg.com
VOICE-MESSAGE: 5/2 (1/0)
```

In this example, the bottom headers provide the device with the message indication. The *MESSAGES-WAITING* header identifies that there is a message waiting for the specified subscriber account. The *MESSAGE-ACCOUNT* identifies the message account that this message is for, and the *VOICE-MESSAGE* header indicates the number of new/old messages (in that order) and the number of new and old urgent messages. In this example, my mailbox is holding five new messages, two old messages, one new urgent message, and no old urgent messages.

Again, the *NOTIFY* message does not identify how to notify the subscriber of these messages. These headers are intended to be interpreted by the device and, depending on the type of device, use whatever method is available. This could mean illuminating a light on the device, or sending a message to the subscriber via text messages.

The following RFCs define the various aspects of event notification:

- Message Summary (RFC 3842)
- Conference (RFC 4575)
- Presence (RFC 3856)
- Reg (RFC 3680)
- Refer (RFC 3515)

Interworking with the PSTN

As mentioned at the beginning of this chapter, it is unlikely that service providers will wake up one morning and simply switch off their switched network and turn on the IP network. The switched network consists of many investments that cannot be simply thrown away. It will take many years to migrate the switched portions of the network to an IP network.

Most service providers have already begun the migration. The typical approach is to begin by deploying an IP backbone for use in transporting packetized voice and data. This IP backbone then allows operators to begin migrating specific portions of their network as they need.

In today's traditional telecom networks, the actual setup and teardown of a voice call takes place through a dialog between two switches. A circuit connection is established between the two switches, and through subsequent switches until a path has been established all the way to the destination. Each switch in the path establishes a dialog and exchanges information regarding the voice call, much as we described in the preceding sections.

The main difference is that the dialog uses a different protocol, Signaling System #7 (SS7), and the dialog is point-to-point. In other words, SS7 is used to connect each individual leg of the call. This means you will see multiple groups of messages for a single call. In SS7 terminology, the first message that is used to initiate a circuit connection is the initial address message (IAM). This is somewhat analogous with the SIP *INVITE*.

The SS7 IAM is sent between each individual switch in the voice path, using digital channels dedicated for the purpose of signaling. The voice transmission itself can be sent on accompanying digital channels, but this is not required. The signaling provides the next switch in the path information needed to establish a connection and route to the next hop in the route.

Figure 4.2 shows the typical call flow for an SS7 network. Note that the *IAM* is sent to the next switch in the route referencing the specific circuit to be used between the two switches. A new *IAM* is then created using much of the same information as the previous *IAM,* but with a new circuit identification. The *IAM* then is used to establish a connection point in the route, and not every connection point end to end.

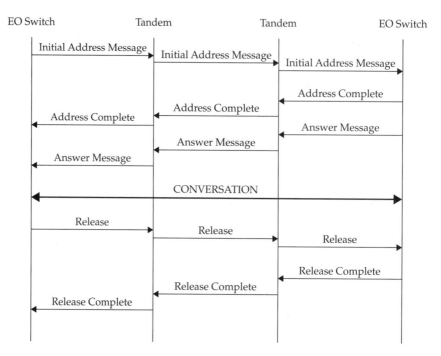

Figure 4.2   SS7 call flow

This is important to understand because this is a big difference between SIP and SS7. A SIP *INVITE* is sent end to end, routed by proxies along the path. There really is no "hard circuit" making the connection along the way hence SIP is not really point-to-point.

Also note that in SS7, each switch in the path is stateful, and is responsible for its own connection, but does not have any knowledge of the end-point state (other than what it receives from its adjacent switch). The only time any switch in the path is aware of a change in state is when it receives a message from an adjacent switch advising it that the connection is to be released.

In the SS7 domain there is no registration process. A device gets connected via hard-wire connections and uses signaling to notify the network when it wants to establish a connection. The phone going off-hook forms a connection path back to the end-office switch, which is the first form of signaling. The digits being dialed create a tone, which is also a form of signaling (providing the destination address).

The SS7 network is part of the Intelligent Network (IN) architecture (this is also sometimes referred to as the AIN). The concept of the IN was to create a separate network that would be used for all call control. This network would consist of routing nodes, called signal transfer points (STPs) responsible for routing signaling messages through the network to the appropriate switching nodes (service switching points, or SSPs).

The network would also have application servers, known as service control points (SCPs). These SCPs would then be used to store routing instructions, calling parties' names and numbers for display purposes, and even number portability databases. Of course these are all in widespread use today.

The problem with the IN is that every switch in the network must be configured with software to be able to access the IN elements. This proved to be quite costly as vendors quickly inflated the price of software to ensure profits. The SCP concept never really fared well outside of the U.S. because it was simply far too costly to implement. The revenues gained through a new service were quickly eroded by the cost of deployment.

Another inherent issue with the IN was its inability to support other media types. The call control protocol used within the IN was of course SS7, which was originally developed for managing voice calls. Later it was adapted to support data as well, and then broadband. However there were issues in supporting video and other forms of media, especially in a switched environment.

The protocol itself is capable of supporting other media types (with modifications), but the network is based on switches and therefore unable to support those media types. Not to mention that the TDM network is not well suited for multimedia. So as IP is deployed in these networks, and operators begin moving voice calls to the IP network, interworking becomes critical.

The central focal point within the SS7 network is the STP. While its primary function is that of a router, it is also a firewall, a gateway, and many other things to the network. My book *Signaling System #7* (McGraw-Hill, 2006) provides much greater detail as to the functions of the STP; our interest here is interworking.

SS7 is well suited for the IP network. After all, SS7 is nothing more than packet data, and as long as a reliable real-time transport can be provided, SS7 can work. The problem with the IP network is its ability to support real-time applications (such as SS0), and therefore a new set of protocols had to be developed to support SS7 and other real-time functions in an IP environment.

The Signaling Transport (SIGTRAN) protocol was developed for just that. Replacing the Message Transfer Part (MTP) in the SS7 network with protocols developed for IP was part of the challenge, since the MTP provides specific network management functions that have to be replicated within the IP domain. The other challenge was developing a transport protocol to replace TCP and UDP for real-time applications. The Stream Control Transport Protocol (SCTP) was then developed to provide this function.

So now that SS7 can be supported in the IP domain, operators have begun moving their SS7 facilities onto IP. This allows them to eliminate all of the TDM facilities supporting signaling and connect their STPs directly to MGCFs. The MGCF provides the connectivity between the SS7 and SIP domains. The MGCF acts as an SSP in the SS7 domain and a SIP user agent in the SIP domain.

The MGCF manages all voice communications within its own domain. This means interworking with its own SS7 network. When connecting to another domain, a break-out gateway control function (BGCF) is used. The BGCF provides additional security functions for managing external networks.

The use of the BGCF also simplifies routing in and out of the network. Without aggregating traffic to some border gateway, other networks would need access to all of the MGCF nodes within a network. This is one of the reasons the traditional switched network always used a hierarchical network architecture.

When a call is placed in the PSTN to a destination in the SIP domain, a telephone number is dialed from the PSTN origination. This telephone number is required for routing within the SS7 domain. The call is routed then to the MGCF within the network according to the telephone number.

The MGCF must then decide the destination of the call; if the call is to be terminated within the SIP domain within the network or if the call is to be routed to another network. If the call is to be routed to another network, the MGCF will forward the SIP *INVITE* to the proper MGCF according to the dialed digits, which have now become a TEL URI. The SS7 message *(IAM)* is converted at the MGCF to a SIP *INVITE*.

To convert the telephone number to a TEL URI, the MGCF must first determine the domain for the TEL URI from the digits dialed. The domain can only be identified by accessing an ENUM function within the network. The ENUM function will provide the proper domain for the TEL URI. The MGCF then must query a DNS to obtain the IP address for routing to the right MGCF/BGCF in the network.

This means two separate queries must be made to determine how to route a call from the PSTN into the SIP domain. The same is true of calls that originate in the SIP network and are terminated (or routed through) the SS7 domain, but the ENUM is converting the TEL URI into an E.164 number. Some operators have discussed putting the ENUM function into their DNS to reduce the number of queries required.

Figure 4.3 shows the call flow for calls originating in the SIP network. The *INVITE* is converted to an SS7 *IAM,* and a 100 TRYING response is sent in the backward direction. The MGCF will then verify that the proper codecs are available at the media gateway (MG), and if so return the response 183 SESSION PROGRESSION.

The MGCF sends the *IAM* toward the terminating switch and waits for the *ACM* to be sent by the switch. When the MGCF receives the *ACM,* it will send the 180 RINGING response to the originator. When the MGCF receives the *ANM* message from the SS7 network indicating that the called party has answered the call, it returns the 200 OK response to the originator. The originator of the call then sends the *ACK* toward the MGCF. There are no equivalent responses to the *ACK* on the SS7 side of the network.

There are many variables that determine what SIP requests/responses are sent and which SS7 messages are sent. Table 4.1 provides mapping between SS7 release cause codes and SIP responses.

TABLE 4.1   ITU SS7 Cause Codes Mapped to SIP Status Codes

SS7 Cause Code	SIP Status Code
1 Unallocated number	404 Not found
2 No route to network	500 Server internal error
3 No route to destination	500 Server internal error
4 Send special information tone	500 Server internal error
8 Preemption	500 Server internal error

TABLE 4.1   ITU SS7 Cause Codes Mapped to SIP Status Codes (*continued*)

SS7 Cause Code	SIP Status Code
9 Preemption circuit reserved for re-use	500 Server internal error
17 User Busy	486 Busy here
18 No user responding	480 Temporarily unavailable
19 No answer from the user	480 Temporarily unavailable
20 Subscriber absent	480 Temporarily unavailable
21 Call rejected	480 Temporarily unavailable
22 Number changed	410 Gone
27 Destination out of order	502 Bad gateway
28 Invalid number format (address incomplete)	484 Address incomplete
29 Facility rejected	500 Server internal error
31 Normal unspecified	480 Temporarily unavailable
34 No circuit/channel available	480 Temporarily unavailable
38 Network out of order	480 Temporarily unavailable
41 Temporary failure	480 Temporarily unavailable
42 Switching equipment congestion	480 Temporarily unavailable
43 Access information discarded	480 Temporarily unavailable
44 Requested circuit/channel not available	480 Temporarily unavailable
46 Precedence call blocked	480 Temporarily unavailable
47 Resource unavailable–unspecified	480 Temporarily unavailable
50 Requested facility not subscribed	500 Server internal error
57 Bearer capability not authorized	500 Server internal error
58 Bearer capability not presently available	500 Server internal error
63 Service option not available, unspecified	500 Server internal error
65 Bearer capability not implemented	500 Server internal error
69 Requested facility not implemented	500 Server internal error
70 Only restricted digital information bearer capability is available	500 Server internal error
79 Service or option not available–unspecified	500 Server internal error
88 Incompatible destination	500 Server internal error
91 Invalid transit network selection	404 Not found
95 Invalid message	500 Server internal error
97 Message type non-existent or not implemented	500 Server internal error
99 Info element/parameter non-existent or not implemented	500 Server internal error
102 Recovery on timer expiry	480 Temporarily unavailable
103 Parameter non-existent or not implemented, passed-on	500 Server internal error
110 Message with unrecognized parameter, discarded	500 Server internal error
111 Protocol error, unspecified	500 Server internal error
127 Interworking, unspecified	480 Temporarily unavailable

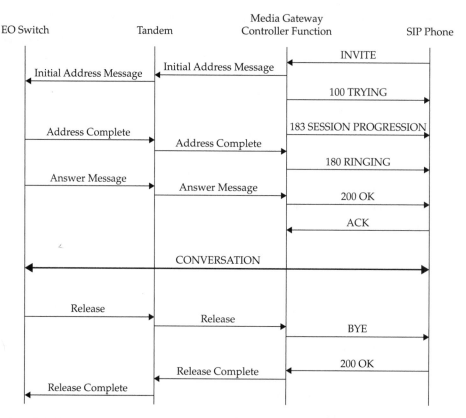

Figure 4.3   Call flow between SS7 and SIP domains

There are a few cause codes that are different for ANSI SS7, as shown in Table 4.2. These cause codes are specific to networks in the U.S and wherever else ANSI SS7 is supported.

TABLE 4.2   ANSI SS7 Cause Codes Mapped to SIP Status Codes

SS7 Cause Code	SIP Status Code
23 Unallocated number	404 Not found
25 Exchange routing error	500 Server internal error
26 Misrouted call to a ported number	404 Not found
45 Preemption	500 Server internal error
46 Precedence	500 Server internal error
51 Call type incompatible with service request	500 Server internal error
54 Call blocked due to group restrictions	500 Server internal error

Of course there is not always a one-for-one mapping between two disparate protocols. Table 4.3 provides SIP status codes mapped to ITU SS7 cause codes.

TABLE 4.3   SIP Status Codes Mapped to SS7 Cause Codes

SIP Status Code	SS7 Cause Code
400 Bad request	127 Interworking
401 Unauthorized	127 Interworking
402 Payment required	127 Interworking
403 Forbidden	127 Interworking
404 Not found	1 Unallocated number
405 Method not allowed	127 Interworking
406 Not acceptable	127 Interworking
407 Proxy authentication required	127 Interworking
408 Request timeout	127 Interworking
410 Gone	22 Number changed
413 Request entity too long	127 Interworking
414 Request-URI too long	127 Interworking
415 Unsupported media type	127 Interworking
416 Unsupported URI scheme	127 Interworking
420 Bad extension	127 Interworking
421 Extension required	127 Interworking
423 Interval too brief	127 Interworking
480 Temporarily unavailable	20 Subscriber absent
481 Call/transaction does not exist	127 Interworking
482 Loop detected	127 Interworking
483 Too many hops	127 Interworking
484 Address incomplete	28 Invalid number format
485 Ambiguous	127 Interworking
486 Busy here	17 User busy
487 Request terminated	127 Interworking
488 Not acceptable here	127 Interworking
491 Request pending	No SS7 equivalent
493 Undecipherable	127 Interworking
500 Server internal error	127 Interworking
501 Not implemented	127 Interworking
502 Bad gateway	127 Interworking
503 Services unavailable	127 Interworking
504 Server timeout	127 Interworking

(*Continued*)

TABLE 4.3   SIP Status Codes Mapped to SS7 Cause Codes (*continued*)

SIP Status Code	SS7 Cause Code
505 Version not supported	127 Interworking
513 Message too large	127 Interworking
580 Precondition failure	127 Interworking
600 Busy everywhere	17 User busy
603 Decline	21 Call rejected
604 Does not exist anywhere	1 Unallocated number
606 Not acceptable	127 Interworking

The status codes, remember, indicate unsuccessful attempts to establish a session, much as the SS7 cause codes indicate failures to establish connections in the PSTN. There is actually a bit more information provided within the SIP status codes, but this is primarily because they cover so many different variables, while SS7 is only concerned with the connection of voice and data circuits.

The actual messages used to establish connections must also be mapped. When a call is originated in the SS7 domain (or in the SIP domain, as we saw in the example in Figure 4.3), the signaling must be mapped so that the MGCF and the SIP proxies understand what must be managed within their own domains.

When a MGCF receives an SS7 *IAM* it will extrapolate specific parameters from the *IAM* and use the values within these parameters for the SIP *INVITE* that it must generate and forward to the SIP domain. The following SS7 parameters are captured by the MGCF for this purpose:

- Called party number
- Calling party number
- Calling party's category
- Forward call indicators
- Hop counter
- Nature of connection indicators
- Transit network selection
- User service information

These parameters provide vital information needed to successfully establish a session within the SIP domain, and therefore the parameters (found in the SS7 *IAM*) are converted to their SIP equivalents and used in the SIP *INVITE*.

Called party number   It wouldn't make sense if this parameter were not necessary, since this is the parameter that carries the destination address. In the PSTN, these would be the digits dialed by the originating party. The digits dialed, of course, are not going to be much use to the SIP proxies, so the MGCF will need to use these digits to query the ENUM application and determine the proper TEL URI.

Calling party number   This parameter would be used for the *P-ASSERTED-IDENTITY* header. The calling party number is crucial to ensuring the originator of the session and is authorized to access the network. The calling party number parameter consists of numerous other parameters, including the *numbering plan indicator* and the *nature of address indicator.*

When a SIP *INVITE* is being converted back to an *IAM* for forwarding into the SS7 network, the *numbering plan indicator* is set to *ISDN (telephony) numbering plan* and the *nature of address indicator* is set to either national or international, based on the TEL URI of the SIP *INVITE*. This means that the MGCF must be able to read and decipher the country code in the SIP TEL URI.

Calling party's category   This is a mandatory parameter for SS7, which means the MGCF must include this parameter when forwarding a call into the PSTN. The value is set to 0000 0000 CALLING PARTY'S CATEGORY UNKNOWN. It can also be set to NS/EP CALL or EMERGENCY SERVICE CALL for emergency calls that should be given high priority treatment in the network, dependent on the network's implementation.

Forward call indicators   The forward call indicators must be set as shown in Table 4.4.

Hop counter   The hop counter is used within SS7 to determine if there is circular routing in the network. The ISUP *IAM* populates this parameter with the value derived from the SIP *INVITE MAX-FORWARDS* header, and vice versa. When the MGCF receives either parameter, it decrements the value by one and forwards the new value in the appropriate header, depending on the direction of the message (toward the SS7 or the SIP domain).

Nature of connection indicators   These are also mandatory parameters and must be set according to Table 4.5.

TABLE 4.4   Forward Call Indicator Values

Bit D	Interworking indicator	Interworking encountered	1
Bit F	BICC/ISUP indicator	BICC/ISUP used all the way	0
Bits HG	BICC/ISUP preference indicator	BICC/ISUP not required all the way	01
Bit I	ISDN access indicator	Originating access non-ISDN	0
Bit M	Ported number translation indicator	Number not translated	0
Bit M	Ported number translation indicator	Number translated	1

TABLE 4.5   Nature of Connection Indicator Values

Bits AB	One satellite circuit in the connection	01
Bits DC	Continuity check not required (ISUP) or no COT to be expected (BICC)	00
	Continuity check performed on a previous circuit (ISUP) or COT to be expected (BICC)	10
Bit E	Outgoing echo control device included	1

Transit network selection   As the SIP message is forwarded to the MGCF, the request-URI is changed to contain the value of the transiting carrier (per normal SIP routing rules). The MGCF then uses this to determine the transit carrier. The Carrier Identification Code (CIC) is used from the request-URI to determine this. Coming from the SS7 side of the network, this parameter may or may not be populated.

User service information   This information is mapped into the SDP portion of the SIP *INVITE* when going from PSTN to SIP. When a call is going from the SIP network to the PSTN, then the *media type, bandwidth,* and *access lines* are mapped from the SDP into the *USI* parameter in the outgoing SS7 *IAM*. There is actually much more detail sent for the USI regarding the session, as detailed in Table 4.6.

TABLE 4.6   SDP Contents Mapped to SS7 USI

SIP SDP			SS7 User Service Information			High-Layer Characteristics
Media Line (m=)	Bandwidth Line (b=)	Access Line (a=)	Info Transfer Rate	Info Transport Capability	User Info Layer 1 Protocol Indicator	High-Layer Characteristics Identification
audio RTP/AVP 0	None or 64 Kbps		64 Kbps	3.1 KHz audio	G.711 µ-law	
audio RTP/AVP	64 Kbps	rtpmap PCMU/8000	64 Kbps	3.1 KHz audio	G.711 µ-law	
audio RTP/AVP 9	AS:64 Kbps	rtpmap:9 G.722/8000	64 Kbps	Unrestricted digital info With Tones and announcements		
audio RTP/AVP	AS: 64 Kbps	rtpmap CLEARMODE/8000	64 Kbps	Unrestricted digital info		
image udptl t38	Up to 64 Kbps		64 Kbps	3.1 KHz audio		Facsimile Group 2/3
image tcptl t38	Up to 64 Kbps		64 Kbps	3.1 KHz audio		Facsimile Group 2/3
audio RTP/AVP	384 Kbps	Rtpmap CLEARMODE/8000	384 Kbps	Unrestricted digital info		
audio RTP/AVP	1472 Kbps	Rtpmap CLEARMODE/8000	1472 Kbps	Unrestricted digital info		
audio RTP/AVP	1536 Kbps	Rtpmap CLEARMODE/8000	1536 Kbps	Unrestricted digital info		

Throughout the session establishment process, the MGCF/BGCF maintains the dialog with the SIP entities, while also managing the SS7 connection points (the MGCF/BGCF should also have a SS7 address, or point code).

Certainly as the IP network continues to expand and other services begin migration to the IP backbone, the need for switched facilities lessens. This is when operators begin migrating all services to the IP backbone under SIP control. Again this is not an overnight function, but something that takes place over a long period of time (possibly 10–15 years). The STPs in the SS7 network migrate to a call session control function (CSCF) as defined by the 3GPP in their IMS architecture, providing a centralized session control function in the core network, while supporting the same level of authorization and authentication realized now.

Establishing a Session in SIP

A session in SIP networks can be anything—a voice call, e-mail, text message, or video stream. This is the beauty of SIP; network operators are able to use SIP to control all forms of communications within their networks, not just voice.

Certainly this is the issue with switched networks today. The technologies and the basic architecture of modern-day networks are not conducive to forms of media other than voice. They were designed and engineered specifically for this purpose. To try to use these same networks for other media types would not be optimal.

IP, on the other hand, supports all forms of media. While initially IP brought about its own issues (such as latency and support for real-time media), these have been overcome, and IP is now a great choice for all networks. That doesn't mean IP does not have its issues and challenges. There are still challenges in implementing an all-IP network, or even adding an IP network to a switched network.

Still, using IP for a transport has proven to be the best option for supporting today's networks. Early on, many different control protocols were used, a fact that has added to many implementation and interworking issues. The industry has begun settling on one protocol now for session control, and that protocol is SIP.

The concept of virtual connections between two entities is not a new concept. You will find many concepts applied in SIP that have existed in many other networks before. SIP is really not all that new, having been derived from the Hypertext Transport Protocol (HTTP) and the Simple Mail Transport Protocol (SMTP).

There are two things that need to happen within a SIP network for connections to be established. The first step is to establish a dialog between the entities trying to connect. This allows them to exchange parameters regarding the connection they wish to make.

We see a similar concept in existing networks today, where switches communicate the parameters of the connections they wish to make with one another. The difference is in the switched network, the dialog is actually the same as the session description. There is no need to establish a dialog separate from the session, since switched networks cannot support multimedia sessions, and therefore there is no need for the switches themselves to exchange information apart from the signaling itself.

Another fundamental difference between legacy signaling methods such as SS7 and SIP is that legacy signaling is used to identify a specific resource and what is required to support the voice transmission on that facility.

SIP is similar in that it describes what is needed to support the session, but it does not concern itself with the facilities. SIP describes what is needed at the end device supporting the session, including what media types the device is going to have to support. The facility is not described in the SIP dialog.

Another fundamental difference is that SIP can support modifications to a session in progress without affecting the session. For example participants can be added or deleted from a session, and media types can be added or deleted as well. This is not possible using signaling such as SS7.

In this section we will look at how SIP establishes a session, how sessions are modified and terminated, and how SIP routing works, whether you are using loose or strict routing.

Accessing the Network

The first step in establishing a session is getting access to the network. While this is really outside the scope of this book, I have included this discussion to clarify what takes place prior to SIP being engaged.

When a device is activated, it must first seek a network connection. On a PC, the network card looks for a connection through either the Ethernet card or the WiFi adapter. If it senses a carrier (and hence a connection) on either one, it exchanges IDs with the network provider to obtain an IP address.

The IP address is typically allocated dynamically. In other words, you are not usually assigned a fixed IP address, but the first IP address available at the time. This avoids using up all of the IP addresses available, and it allows you to change locations without having to be concerned about your IP address. In fact, the network has the ability to discover your IP address as you begin using the network.

The Dynamic Host Configuration Protocol (DHCP) is used to obtain an IP address from the network server. This is done over the IP network. Once you have obtained your IP address, you can then begin using applications on our device for communications. The applications you use will determine what protocols are used for session control. For example, when you launch your Web browser, you are using HTTP to access Web sites.

If you launch a communications application that is SIP enabled, you will immediately begin using SIP to set up sessions and communicate with other users. The IP address that you obtained from the network stays assigned to your device as long as you maintain the connection.

In some cases, however, you may move from one location to another. This is where wireless comes into play. WiFi does not support roaming, so when you wander outside the range of your access point, you lose your connection (as well as your IP address).

WiMax, on the other hand, does support roaming, so as you move from one location to another, your IP address is re-assigned based on availability, but your connection is not lost. The connection is maintained throughout the "handoff." This is the same concept used in cellular networks.

The IP connection does not connect you to any one network necessarily. It may be just a public network you are connecting into, so you may have to connect into your local service provider network to be able to use services. For example, you may have an account with Google, in which case you will need to log on to the Google network to be able to access your account and use their services.

Typically, if you are connecting into a VoIP network, you will connect with a local media gateway controller (MGC) that services your area. The MGC is able to communicate with devices via SIP and may even use SIP to communicate to other MGCs. It is my own personal viewpoint that all networks will migrate to SIP and may also even implement the IMS as an architecture for implementing their SIP networks.

The IMS architecture is defined by the 3GPP for implementing SIP in a communications network. It calls for call session controllers deployed in such a fashion as to support many of the processes and procedures already being supported in GSM networks today. The IMS is defined in my other book, *The IP Multimedia Subsystem: Session Control and Other Network Operations* (McGraw-Hill, 2007).

Even with the IMS, the first step is accessing an IP network and obtaining an IP address. This means that "someone" (the local telephone company) is going to have to be in the business of providing these connections to our homes and businesses.

The connection is maintained as long as the device is active. However, the connection does not mean any sessions are established. Sessions are only established when your device or any other device wishes to communicate with another device.

When devices are ready to communicate, they must form a virtual connection. This is referred to as a *dialog* in SIP terms. The dialog allows the devices to exchange data between one another and is maintained as described in the next section.

Initiating a Dialog

SIP uses an offer/answer mechanism for establishing a session. This means that any entity wishing to establish a connection with another entity must first initiate an offer (the offer contains the parameters for establishing the connection, such as resources required). The receiving entity must then answer the offer, either accepting or rejecting the offer.

The user agent client (UAC) function within the device is responsible for initiating an offer. The UAC, remember, is a function resident in all network entities capable of establishing and participating in sessions. The offer is made to the user agent server (UAS) function at the destination device.

If the UAS is going to accept the offer, it will send an answer back to the UAC, which will then return an acknowledgment to the UAS. The final act of sending the acknowledgment is what establishes the session.

A session is a logical connection between two entities wishing to communicate. There may be multiple devices involved in any one session. For example, in a conference call there are multiple recipients of a session offer. Each of the recipients must send its own answer to the offer. Each of the answers must be managed separately to maintain the status of each of the participating entities.

The offer must also contain details about the session itself. These details include any resources such as codecs that must be provided and media types to be supported. The details may also include addresses where certain aspects of the session can be learned. This information is contained within the Session Description Protocol (SDP) carried by the offer.

To communicate the various aspects of a session with one another, the participating entities must establish a unique dialog. This is also a virtual connection between each of the recipients and the originator of the offer. The originator of the offer will maintain a separate dialog with each of the participants.

When two devices communicate with one another, they exchange a set of messages (referred to as transactions). For example, a cell phone originating a call to another cell phone would send an *INVITE* to the other phone requesting a connection be made. The recipient of the *INVITE* will determine whether to accept or reject the transaction.

If the cell phone chooses to accept the invitation to a session, then it will send a response to the request and exchange other messages entering into a dialog with the other device. The dialog then becomes a logical connection between the communicating entities for the purpose of exchanging SIP messages regarding a session. Only *INVITE*, *SUBSCRIBE*, and *REFER* will lead to the creation of a dialog.

Each user agent establishes its own dialog with the user agent client (the requestor). The dialog is then used by the user agent to maintain the status of the dialog and associated sessions. The session is not the same as a dialog, since a session can involve multiple user agents communicating with one user agent client.

In other words, the dialog is established between each entity involved in the same session, as depicted in Figure 5.1. The illustration shows the user agent client with a dialog established with multiple user agent servers, exchanging session control information for the same session (as would be the case in a conference call).

As can be seen in the figure, the user agent client can now delineate transactions between each of the entities participating in the same session. This allows the UAC to correlate responses from each individual UAS and treat each one independently even though they are all participating in the same session.

Each dialog requires a dialog ID, which is derived from the SIP headers. When the UAC sends a request for a session, it will expect a response. In that response will be required headers (as we will discuss in a moment). The response must contain the *TO*, *FROM*, and *CALL-ID* headers. The *TO* and the *FROM* headers will include the *TAG* parameter. The *TAG* parameter is used by each user agent for correlating requests with responses, but it is also used for calculating the dialog ID.

The dialog ID then becomes unique for the UAC and each of the UASs. In other words, the UAC creates its own dialog ID, while each of the UASs will create its own unique dialog ID. This is not communicated to other entities, since the use of the dialog ID is local (used by the user agent).

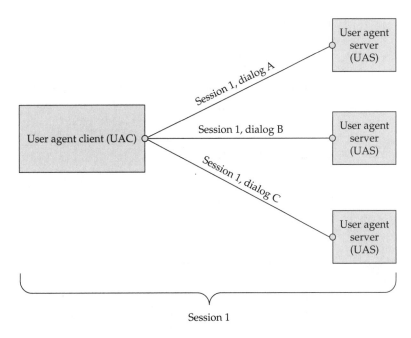

Figure 5.1   The difference between a dialog and a session

Note that a dialog ID cannot be determined until the UAC receives a 2*xx* response back from each of the UASs. This is because the *TO* header does not actually contain the *TAG* parameter until the response is generated by the UAS. Remember that the *TAG* is used by the UAS for correlation of responses (more on that when we discuss the various message headers and their parameters).

So in summary, a UAC will generate a SIP message and send it to one or more UASs. Each UAS will then answer the UAC by sending a response message. The response message contains at a minimum the *TO*, *FROM*, and *CALL-ID* headers, which are used in creating the dialog ID. Once all of these are done, the dialog is considered established, as depicted in the call flow shown in Figure 5.2.

Figure 5.2 also shows a handshake. Once the dialog has been established and the acknowledgment is received, the handshake sequence between two entities has been completed. This marks the beginning of a session where bearer traffic is exchanged. A session cannot begin without completion of a handshake between the two entities. Think of the handshake as an agreement between all involved user agents to engage in the transfer of bearer traffic.

To establish a session, a sequence of messages must be exchanged between all involved parties. This sequence of messages consists of an offer and an answer. The offer is established by the user agent client wishing to begin a session. It must contain all of the data necessary to establish the session and transfer bearer traffic.

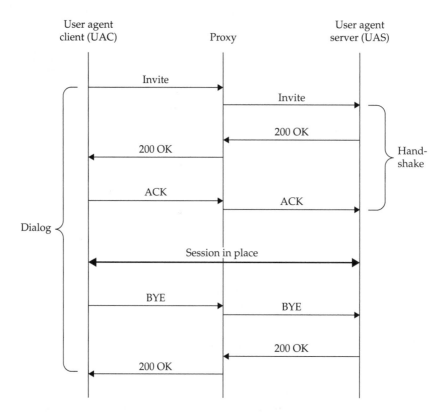

Figure 5.2 Establishment of a dialog

The offer therefore carries the description of the session, including the encoding methods, required media support, and any other support required for the session at the receiving end. The offer may be contained in the initial request, or else it may be carried in a subsequent response.

The offer is then followed by an answer. Each entity involved in the session must provide an answer to the offer. The answer in essence is accepting the offer and ac-knowledging it can support the requirements specified in the offer. If no requirements were specified in the offer, the answer may contain requirements.

Once an offer has been made, another offer cannot be made to the same destination until an answer has been received for the first offer. The answer can be acceptance or rejection of the offer, but multiple offers are not allowed to one destination without answers being completed.

The offer and the answer use specific methods. A method in SIP parlance is a mes-sage type. There are two forms of SIP messaging: the SIP request and the SIP response. These are explained in more detail in the sections that follow.

Client Request

The UAC is responsible for initiating all requests. Remember that all devices will support both the client and server functions. The client function originates requests, and the server function responds to these requests.

The request for a session typically consists of sending an *INVITE*. However, requests are not limited to this method. *REGISTER* is also a request, as is *MESSAGE*, *OPTIONS*, and *CANCEL*. It all depends on the context of the method. All methods are considered requests, while the responses are identified by a number. So it is safe to say that the UAC uses a method (or SIP message type) to communicate a request, and the UAS uses one of the six classes of responses to answer the request.

Each request begins with a start line as described in Chapter 2, identifying the method being sent (and therefore the nature of the request). We can also assume that certain methods are used for specific types of requests. For example, if we review the various methods again, these methods assume a special importance:

- *INVITE*   Used to request a virtual connection between two or more entities to exchange user data (voice calls, data, e-mail, etc.). This method is used to set up a session between the identified entities.

- *BYE*   Used to release a session in progress. Requires the session be established (and a dialog established).

- *ACK*   Used to complete the handshake process and establish a session.

- *CANCEL*   Request that a request previously sent be canceled. This requires that a session not be established, since that would require the use of the *BYE* method.

- *REGISTER*   Request to register the location of the device in the network (current IP address, etc.).

- *OPTIONS*   Request to identify the resources available at a server location, such as an application server.

- *MESSAGE*   Request to accept a text message.

The UAC should also include within the request the requirements for the session. This means identifying the resources that will be required to support the session, including any media resources needed (such as voice or video). In the event that the request does not contain any session description, then the UAS has the option of sending a session description identifying what it can or is willing to support.

The session description is carried in the message body of the request using the Session Description Protocol (SDP) in most cases, although this is not a requirement. The SDP was developed for this purpose in any type of network and is certainly not limited to SIP. In other words, there is no dependency on SIP for SDP, or vice versa.

When the UAC sends a request, it adds the header *FROM*, with a TAG value. The TAG is assigned by the UAC as a unique identifier and is used by the UAC to track the responses to the request. In other words, when the UAC receives a response to its request, the TAG value will be the same value in the response, allowing the UAC to correlate responses to requests. There are other identifiers though, as you will see.

The UAS will add its own TAG value to the *TO* header, which is used by the UAS to correlate requests and responses. However, both entities will also use the dialog ID for correlation and management of the dialog between the two entities. The dialog ID as we discussed earlier is calculated using both TAG values and the *CALL-ID* header value. This then allows the UAC and the UAS to correlate requests/responses for specific sessions.

The *ACK* is sent by the UAC when it receives a final response from the UAS. The final response does not necessarily have to be acceptance of the session. The *ACK* will contain a *BRANCH* identifier found in the *VIA* header. This is used by the stateful proxies in the path of the request so that they can correlate requests and responses.

This is necessary because proxies may send responses (such as provisional responses) back to the UAC. The *BRANCH* field is used by these proxies to correlate requests and responses. The proxy adds the *BRANCH* value to the *VIA* header as the message is forwarded by the proxy; therefore, if there are multiple stateful proxies, there will be multiple *BRANCH* fields. The *VIA* header also contains the address of the proxy.

The UAC may continue sending a request until it receives a response. Each UAC starts an internal timer, and when that timer expires, the UAC resends the request. This continues until the UAC receives a response from the UAS. We will discuss responses in more detail in the next session, but this is important to remember because retransmission of requests could have adverse effects on the network if provisions are not made for this.

The UAC may also send a single request to multiple addresses. This could be the case for a broadcast, for example. Forking proxies could also be used to distribute requests to multiple destinations. When this occurs, the forking proxy is responsible for managing the responses from each of the UASs. The request is received by the proxy, which then forks the response to other destinations according to its own internal routing instructions.

The UAC, therefore, is not aware of the additional addresses that the request may have been sent to, so the forking proxy must maintain the state of the request in association with each individual UAS. This, again, is the purpose of attaching a *BRANCH* field to the *VIA* header when the message is received by the proxy. The proxy is then able to use this value to correlate responses with the request that it forks.

The UAC addresses the request according to the destination public address. This address is contained, then, in the request-URI found in the start line of the request. Note that the request-URI is not always the final destination of the request. It can also be the next hop within the network, depending on the implementation. We will discuss routing in the SIP network a little later on.

The request-URI is critical to routing the request, because not only does it identify the address of the recipient (or the next hop in the network) but it also identifies the domain to which the request is to be routed. This means that the first proxy to receive the request must determine how to route the request from the domain name in the address. The proxy, therefore, will query DNS to find the address for the access point within the domain.

Now that we have discussed the creation of a request, let's look at how a UAS responds to a request.

Server Response

There are basically six types of responses as you saw in Chapter 3; provisional, successful, redirection, client error, server error, and global error. The use of numbering simplifies the use of these responses and allows vendors to use the numbering scheme within their applications to make applications more user friendly.

When a request is sent by the UAC, a timer is sent at the UAC to ensure a response is received within a given time. The default for this timer is typically three seconds. If there is no response within three seconds, the UAC will retransmit the request, and the timer is reset.

This could pose a problem in large networks, where latency may cause additional delays in the routing or requests. To prevent large volumes of requests from clogging the network, proxies use provisional responses to stop the expiration of the response timer in the UAC. The provisional response extends the time for a response, and it continues to be sent until the final response is received by the UAC. The proxy maintains the state of the request/response and ceases the transmission of provisional responses when it receives the final response for forwarding to the UAC.

The provisional response does not establish a session or a dialog. It simply prevents retransmission of the request by confirming that the request was received by the proxy and is being processed. When the request is forwarded to the next proxy, a timer in the first proxy is set in the same fashion as in the UAC. The first proxy will then wait three seconds before retransmitting the request. The second proxy will therefore send a provisional response to the first proxy. The UAS can send as many provisional responses as it likes, as long as the *CALL-ID* correlates to the initial request.

If there is no response from the UAS, and the response timer in the UAC expires, the UAC will send response 408 Request Timeout toward the UAS. Each of the proxies in the path will also process the response and consider the request as canceled.

This continues throughout the network, until the final response is received by each of the network entities and the UAC. One could then consider SIP networks implemented in this fashion as point-to-point networks, as each node within the network manages the connection to the next hop in the call path.

When the UAS receives the request, it can respond with a successful response or an error response. It can also send a provisional response to delay the retransmission of the request. The remote target URI is replaced with the URI found in the *CONTACT* header of the 2*xx* response if loose routing is implemented. If strict routing is used, then the response is sent in the same path as the request. You'll learn more about routing in the section "SIP Routing" later in this chapter.

When the UAS receives the request, it will compute the dialog ID to determine if this is for a dialog already in progress. If the request is for a dialog already in progress, then the request is treated as a mid-dialog request.

If the UAS does not recognize the dialog ID, it assumes that another UAS has failed and the receiving UAS is providing backup to the primary (intended) server.

This could have been caused by a crash or reboot of the intended UAS. For this reason the UAS will not necessarily reject the request.

The first phase described here is session negotiation. The UAS then examines the SDP to determine if it is able to support the requested session. If the UAS determines that it cannot support the session, it returns a response of 488 NOT ACCEPTABLE HERE. This indicates that the UAS is unable to support the media identified in the SDP.

The UAS is not rejecting the session request in its entirety, but rather it is declining to support for the specified resource. If the UAS were to reject the entire session for other reasons, it would send the response 606 NOT ACCEPTABLE.

The UAC can then change the required resources for the session or abandon the request altogether by sending the *BYE* message. Changing the required resources would require sending a subsequent *INVITE* with new requirements.

So far we have described autonomous responses—those sent by the UAC and UAS functions without user intervention. The UAS also has the option of sending an alert to the end user of the device to allow the user to accept or reject a session request. This of course is managed by the application receiving the request.

For example a user may receive a request to join an instant message (IM). The application receiving the *INVITE* would send a provisional response (such as 180 RINGING) to the UAC while sending an alert and display it to the user. The user must then determine if he or she wants to accept the IM or reject it.

If the user decides to accept the IM, the UAS function within their device would then return a 200 OK response to the UAC. However, if the user decides to reject the request, the UAS would return the response 603 DECLINE.

The entire sequence of messages might look something like this example:

```
INVITE sip:travis.russell@tcg.com sip/2.0
VIA: SIP/2.0/UDP  pchome101@aol.com:5060;branch=z9hG4bK74gh5
FROM: Deby Russell <sip:deby.russell@aol.com>;tag=9hz34567sl
TO: Travis Russell travis.russell@tcg.com
MAX FORWARDS: 70
CALL-ID: 82167534@126.18.27.0
CSeq: 1 INVITE
CONTACT: Deby Russell <sip:deby@126.18.27.0>
CONTENT-TYPE: application/SDP
CONTENT-LENGTH: 154
```

In the preceding message, the request is sent to the user. Notice that a dialog has not yet been established (the *TO TAG* value has not been appended yet).

```
SIP/2.0 200 OK
VIA: SIP/2.0/UDP pchome101@aol.com:5060; branch=z9hG4bK74gh5
FROM: Deby Russell <sip:deby.russell@aol.com>;tag=9hz34567sl
TO: Travis Russell <sip:travis.russell@tcg.com>;tag=1df789jkf
MAX FORWARDS: 70
CALL-ID: 82167534@126.18.27.0
CSeq: 1 INVITE
```

```
CONTACT: Travis Russell <sip:travis.russell@135.18.10.10>
CONTENT-TYPE: application/SDP
CONTENT-LENGTH: 154
```

The UAS responds to the request by sending the 200 OK response as shown. Note that the UAS has now appended the *TO* header with the *TAG* field. The dialog ID can now be calculated using the TAG values from the *FROM* and *TO* headers, as well as the *CALL-ID* value.

```
ACK sip:travis@135.18.10.10 SIP/2.0
VIA: SIP/2.0/UDP pchome101@aol.com:5060;branch=z9hG4bK74gh5
FROM: Deby Russell <sip:deby.russell@aol.com>;tag=9hz34567sl
TO: Travis Russell <sip:travis.russell@tcg.com>;tag=1df789jkf
MAX FORWARDS: 70
CALL-ID: 82167534@126.18.27.0
CSeq: 1 INVITE
CONTACT: Deby Russell <sip:deby.russell@128.18.27.0>
```

The *ACK* can now be sent, establishing the session. Note that the *BRANCH* field in the *ACK* is the same value as contained in the initial *INVITE*. This is so stateful proxies will be able to correlate the requests and responses for this session.

The *BRANCH* value will always start with the same first seven digits. In this case, the first digits are z9hG4bk, indicating that this message was processed using the processes and procedures defined in RFC 3261 SIP 2.0.

One last note about acceptance of a request: The *CSeq* header is used to ensure messages are received in proper sequence. For this reason, the UAS will always verify that the *CSeq* value is the expected value for a dialog. When the initial request is sent, the value is always the lowest number. Each subsequent message must be the next value greater than the initial request. If not, the message is considered out of sequence and the UAS responds with 500 SERVER INTERNAL ERROR.

There can be gaps between received messages, which is allowed. This could be the case, for example, when a proxy challenges a UAC prior to forwarding the request to the UAS. The UAC sends another *INVITE* containing the proper credentials, and with an incremental *CSeq* value.

This provides a simple explanation of session establishment. There are many more different possibilities for session establishment, depending on the method used; however, they all use the same basic framework described here. What we have not discussed yet is emergency session initiation (such as the case for an emergency call to 911 or 0911).

Emergency Session Establishment

For routing purposes, each network must maintain a routing database identifying the proper destination address for all emergency access points (in the U.S. these are referred to as Public Safety Answering Points, or PSAPs).

The mechanism defined is quite simple from a SIP perspective. A stateful proxy within the network is used to maintain and store all of the emergency numbers and URIs for a given area or network. Obviously this is best distributed throughout the network according to geography so that there will be several stateful proxies assigned in accordance with their geography and network topology.

The role of the proxy in this case is to route all SIP requests with a TEL URI of 911 or other similar emergency number to the appropriate destination. The proxy is managed and configured by the network administrator.

Work is ongoing in the U.S. today to redefine emergency network priority service to replace the present Government Emergency Telephone System (GETS). This activity involves the nation's leading telecom providers and defines how SIP calls will be managed for first responders. The outcome of this work will result in the replacement of priority services for government services and military usage.

This means features found in many military networks such as precedence will continue to be supported in the SIP domain. However, modifications and extensions must be defined prior to this being implemented and accepted.

The Department of Defense (DoD) is also defining standards for SIP implementation within military networks, but this topic is outside the scope of this book. Both of these activities are mentioned here to identify that there is work being done for SIP implementation that is outside the scope of RFC 3261, which defines a simple framework for SIP implementation.

SIP Routing

Now that we have covered establishing a SIP session, let's look at how SIP requests and responses are routed through the network. A unique feature of SIP is its ability to locate a subscriber anywhere in the network. This form of mobility is something that is lacking in switched networks, and what has helped increase the popularity of IP-based telephony.

Mobility is an important capability in modern communications, allowing users to receive their phone calls, e-mails, and other communications no matter where they are, as long as they have a network connection. With the advent of 3G/4G wireless and WiFi/WiMax wireless, users can now be connected most anyplace.

Since many of us have multiple devices, we also have multiple identities (as discussed in Chapter 2). Each of our devices carries an identity, which is important when routing because these addresses must be located for proper routing. This is why SIP uses the concept of URIs, which can be resolved to an actual IP address using the Domain Name Server (DNS).

The DNS resolves the URI found in the request-URI of the message into an actual IP address that can then be used for routing to an actual device. If a TEL URI is used, then the ENUM function is queried to resolve the TEL URI to a domain name, which is then resolved by the DNS into an IP address.

So you can see that from the offset there are a number of messages traversing the network just to determine how to route a SIP message to its destination. However, this whole concept of the URI is what enables mobility, since users can change their locations at any time, and hence change their physical addresses anytime, without losing the ability to be reached by other users.

In the next example you see the method *INVITE* with the request-URI in bold. The domain *tcg.com* has to be resolved into an actual IP address at some point, to reach its destination.

```
INVITE sip:travis.russell@tcg.com sip/2.0
```

The proxy is used for routing messages within a SIP network. The proxy can also provide a redirect service, where it provides alternative addresses for reaching a subscriber device. These alternative addresses are stored by the registrar function, which is responsible for receiving a subscriber registration in the network and storing that information for other proxies. The registrar can then provide this function of redirect.

The proxy can operate in two modes; stateless and stateful. A stateless proxy simply routes messages received and erases any information regarding the message. This is the simplest form of proxy, and also the most insecure of implementations (we will explain security in Chapter 7).

The stateless proxy maintains a record of each transaction it receives, including the state of the session. This means also recording requests that it forwards so that it knows where to send responses. Typically with stateful proxies, responses are sent using the same path as the request. This is so that the stateful proxy will be able to maintain the state of a dialog and its associated sessions.

The stateful proxy also acts as a UAS, returning provisional responses and forwarding requests on to their final destinations. All forking proxies are stateful proxies.

There are a number of different means for determining proper routing for a message, but we will focus on two defined methods: loose routing and strict routing. Loose routing makes routing decisions based on available routes and network conditions. The Internet is a good example of where loose routing is used. There is no dedicated path for a message to be routed.

This form of routing is acceptable for most forms of communications, but when voice and other real-time services are provided, there are good reasons to implement strict routing. Strict routing is not as favorable because there are caveats to using it. Strict routing will use deterministic routing for the registration, but all subsequent requests and responses must use the same path recorded during registration.

Loose Routing

Loose routing is always the most favorable from a pure network management perspective, because it uses the resources available with the most economical means. Messages are routed according to available routes and the state of each of those routes. The proxies in the path of the messages become simple routers and do not concern themselves with the state of a session or a subscriber.

However, loose routing means that messages for a session are routed using many different routes, making it difficult to correlate messages for a single session. It also

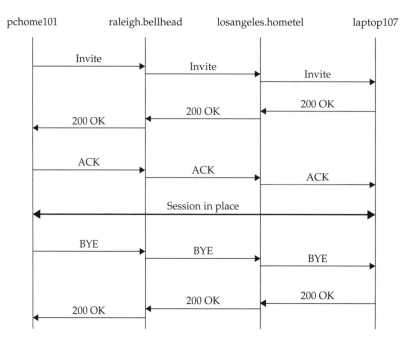

Figure 5.3 Routing example

makes it very difficult to support law enforcement agencies, when looking for call data for a specific session. Identifying the route for the signaling messages and capturing the call data could be tricky in these cases.

The *lr* parameter indicates that loose routing is being used. When loose routing is being used, the request-URI remains constant (instead of changing for each hop). The *TO* and the *FROM* headers never change either, regardless of the routing method used. These two headers are never used for routing in the SIP network.

The *VIA* header is appended by each proxy as the message is routed through the network. The purpose of the *VIA* header is loop detection. If a proxy finds its own address in the *VIA* header, it then knows that this message has already been processed, and the message is looping around the network. This is the only header with addresses that gets appended as the message is routed. Strict routing uses additional headers for recording a route, as we will discuss in the next section.

Routing responsibility for the message is left to each proxy along the route. Any single router is only responsible, however, for its leg of the path, and not end-to-end routing. This is not unlike other signaling protocols such as Signaling System #7 (SS7).

Figure 5.3 shows routing in a SIP network between a UAC and a UAS. The provisional responses have been excluded for simplicity. Examples follow of what each of the messages in this figure would look like when loose routing is used. The final *BYE* and 200 OK have been excluded from these examples:

```
INVITE  sip:travis.russell@tcg.com  sip/2.0
VIA: SIP/2.0/UDP  pchome101@aol.com:5060;branch=z9hG4bK74gh5
```

```
FROM: Deby Russell <sip:deby.russell@aol.com>;tag=9hz34567sl
TO: Travis Russell <sip:travis.russell@tcg.com>
MAX-FORWARDS: 70
CALL-ID: 82167534@126.18.27.0
CSeq: 10001 INVITE
CONTACT: Deby Russell <sip:deby@126.18.27.0>
CONTENT-TYPE: application/sdp
CONTENT LENGTH: 154
```

The request is sent from the UAC to the first proxy, which will then append its own address in the *VIA* header and forward the request to the final destination, which is identified in the request-URI.

```
INVITE  sip:travis.russell@tcg.com  sip/2.0
VIA: SIP/2.0/UDP  raleigh.bellhead.com;branch=z9hG4bK32ad1
VIA: SIP/2.0/UDP  pchome101@aol.com:5060;branch=z9hG4bK74gh5
FROM: Deby Russell <sip:deby.russell@aol.com>;tag=9hz34567sl
TO: Travis Russell <sip:travis.russell@tcg.com>
MAX-FORWARDS: 69
CALL-ID: 82167534@126.18.27.0
CSeq: 10001 INVITE
CONTACT: Deby Russell <sip:deby@126.18.27.0>
CONTENT-TYPE: application/sdp
CONTENT LENGTH: 154
```

Note that there is no *RECORD-ROUTE* header, and that the request-URI does not change. The proxies simply forward the message according to the request-URI address using any available route.

```
INVITE  sip:travis.russell@tcg.com  sip/2.0
VIA: SIP/2.0/UDP  losangeles.hometel.com;branch=z9hG4bK42al3
VIA: SIP/2.0/UDP  raleigh.bellhead.com;branch=z9hG4bK32ad1
VIA: SIP/2.0/UDP  pchome101@aol.com:5060;branch=z9hG4bK74gh5
FROM: Deby Russell <sip:deby.russell@aol.com>;tag=9hz34567sl
TO: Travis Russell <sip:travis.russell@tcg.com>
MAX-FORWARDS: 68
CALL-ID: 82167534@126.18.27.0
CSeq: 10001 INVITE
CONTACT: Deby Russell <sip:deby@126.18.27.0>
CONTENT-TYPE: application/sdp
CONTENT LENGTH: 154
```

The UAS then sends a 200 OK response back to the UAC using the first available route. It does not necessarily use the same route used to reach the UAS.

```
SIP/2.0  200 OK
VIA: SIP/2.0/UDP  laptop107@tcg.com;branch=z9hG4bK32de5
FROM: Deby Russell <sip:deby.russell@aol.com>;tag=9hz34567sl
TO: Travis Russell <sip:travis.russell@tcg.com>;tag=1df789jkf
MAX-FORWARDS: 70
CALL-ID: 82167534@126.18.27.0
CSeq: 1001 INVITE
CONTACT: Travis Russell <sip:travis.russell@135.18.10.10>
```

This is then forwarded on to the next proxy, which will then append its address in the *VIA* header.

```
SIP/2.0  200 OK
VIA: SIP/2.0/UDP  losangeles.hometel.com;branch=z9hG4bK32de5
VIA: SIP/2.0/UDP  laptop107@tcg.com;branch=z9hG4bK32de5
FROM: Deby Russell <sip:deby.russell@aol.com>;tag=9hz34567sl
TO: Travis Russell <sip:travis.russell@tcg.com>;tag=1df789jkf
MAX-FORWARDS: 69
CALL-ID: 82167534@126.18.27.0
CSeq: 1001 INVITE
CONTACT: Travis Russell <sip:travis.russell@135.18.10.10>
```

The final 200 OK response is sent to the UAC through the final proxy.

```
SIP/2.0  200 OK
VIA: SIP/2.0/UDP  raleigh.bellhead.com;branch=z9hG4bK32ad1
VIA: SIP/2.0/UDP  losangeles.hometel.com;branch=z9hG4bK32de5
VIA: SIP/2.0/UDP  laptop107@tcg.com;branch=z9hG4bK32de5
FROM: Deby Russell <sip:deby.russell@aol.com>;tag=9hz34567sl
TO: Travis Russell <sip:travis.russell@tcg.com>;tag=1df789jkf
MAX-FORWARDS: 68
CALL-ID: 82167534@126.18.27.0
CSeq: 1001 INVITE
CONTACT: Travis Russell <sip:travis.russell@135.18.10.10>
```

The *CONTACT* header is used for routing the responses back so that the same proxies do not have to be used.

```
ACK  sip:travis.russell@tcg.com  SIP/2.0
VIA: SIP/2.0/UDP  pchome101@aol.com:5060;branch=z9hG4bK713a2
FROM: Deby Russell <sip:deby.russell@aol.com>;tag=9hz34567sl
TO: Travis Russell <sip:travis.russell@tcg.com>;tag=1df789jkf
MAX-FORWARDS: 70
CALL-ID: 82167534@126.18.27.0
CSeq: 1001 ACK
CONTACT: Deby Russell <sip:deby.russell@126.18.27.0>
```

The *ACK* can now be routed to the UAS as shown in the following two message examples:

```
ACK  sip:travis.russell@tcg.com  SIP/2.0
VIA: SIP/2.0/UDP  raleigh.bellhead.com;branch=z9hG4bK32ad1
VIA: SIP/2.0/UDP  pchome101@aol.com:5060;branch=z9hG4bK713a2
FROM: Deby Russell <sip:deby.russell@aol.com>;tag=9hz34567sl
TO: Travis Russell <sip:travis.russell@tcg.com>;tag=1df789jkf
MAX-FORWARDS: 69
CALL-ID: 82167534@126.18.27.0
CSeq: 1001 ACK
CONTACT: Deby Russell <sip:deby.russell@126.18.27.0>
```

This continues until the *ACK* is received at the UAS. Once the *ACK* is received by the UAS, the session can begin. The final *ACK* in this example would look like this example:

```
ACK  sip:travis.russell@tcg.com  SIP/2.0
VIA: SIP/2.0/UDP  losangeles.hometel.com;branch=z9hG4bK32de5
VIA: SIP/2.0/UDP  raleigh.bellhead.com;branch=z9hG4bK32ad1
VIA: SIP/2.0/UDP  pchome101@aol.com:5060;branch=z9hG4bK713a2
FROM: Deby Russell <sip:deby.russell@aol.com>;tag=9hz34567sl
TO: Travis Russell <sip:travis.russell@tcg.com>;tag=1df789jkf
MAX-FORWARDS: 68
CALL-ID: 82167534@126.18.27.0
CSeq: 1001 ACK
CONTACT: Deby Russell <sip:deby.russell@126.18.27.0>
```

The session is then terminated using the *BYE* method as shown. Notice that the request-URI contains the IP address of the recipient. This comes from the *CONTACT* header and allows the message to be sent directly without having to use the proxies, even though in our example we are still routing to the proxies.

```
BYE  sip:travis.russell@135.18.10.10  SIP/2.0
VIA: SIP/2.0/UDP  pchome101@aol.com:5060;branch=z9hG4bK713a2
FROM: Deby Russell <sip:deby.russell@aol.com>;tag=9hz34567sl
TO: Travis Russell <sip:travis.russell@tcg.com>;tag=1df789jkf
MAX-FORWARDS: 70
CALL-ID: 82167534@126.18.27.0
CSeq: 1001 BYE
CONTACT: Deby Russell <sip:deby.russell@126.18.27.0>
```

```
BYE  sip:travis.russell@135.18.10.10  SIP/2.0
VIA: SIP/2.0/UDP  raleigh.bellhead.com;branch=z9hG4bK32ad1
VIA: SIP/2.0/UDP  pchome101@aol.com:5060;branch=z9hG4bK713a2
FROM: Deby Russell <sip:deby.russell@aol.com>;tag=9hz34567sl
TO: Travis Russell <sip:travis.russell@tcg.com>;tag=1df789jkf
MAX-FORWARDS: 69
CALL-ID: 82167534@126.18.27.0
CSeq: 1001 BYE
CONTACT: Deby Russell <sip:deby.russell@126.18.27.0>
```

```
BYE  sip:travis.russell@135.18.10.10  SIP/2.0
VIA: SIP/2.0/UDP  losangeles.hometel.com;branch=z9hG4bK32de5
VIA: SIP/2.0/UDP  raleigh.bellhead.com;branch=z9hG4bK32ad1
VIA: SIP/2.0/UDP  pchome101@aol.com:5060;branch=z9hG4bK713a2
FROM: Deby Russell <sip:deby.russell@aol.com>;tag=9hz34567sl
TO: Travis Russell <sip:travis.russell@tcg.com>;tag=1df789jkf
MAX-FORWARDS: 68
CALL-ID: 82167534@126.18.27.0
CSeq: 1001 BYE
CONTACT: Deby Russell <sip:deby.russell@126.18.27.0>
```

Note that the *BYE* message contains the tags so that the receiving UA can calculate the dialog ID and determine which dialog/session is to be terminated. Without this the UA will not be able to determine which session it applies to.

Strict Routing

Strict routing can cause issues with congestion over some routes, and it certainly creates imbalance in the load of some parts of the network. However, if this can be overcome (by the configuration of the network and additional bandwidth, for example), then there are many advantages to the use of strict routing. Security is certainly at the top of the list of advantages, but we will discuss that in Chapter 7.

Strict routing relies on a route list for each subscriber. The route list is created during registration. As the *REGISTER* message is routed through the network, each proxy will append the message with its own address using the *RECORD-ROUTE* header. The headers are added to the message in order so that the path can be established.

The route recorded when registering a location is stored at the registrar and becomes a part of the subscriber's registration record. This route list is then shared with any stateful proxy. The route is shared through the use of *ROUTE* headers in the messages. The *ROUTE* messages are added to each request either by the UAC or by a stateful proxy.

For example, some stateful proxies (such as in the IMS) will save the *RECORD-ROUTE* headers for a subscriber when they register, and then use these as a route list for the subscriber for the lifetime of the registration. The route list is then erased and replaced with a new route list when the subscriber registers a new location.

Likewise, when a request is sent to a UAS, the UAS saves the *RECORD-ROUTE* headers as a route list for the subscriber for the duration of the session. There is no need for the UAS to store these addresses once the session is released.

When the UAC has received all of the responses for a request, it then converts the *RECORD-ROUTE* headers into a *RECORD* header (or headers) for subsequent requests. For example, the *ACK* would use the *RECORD* header.

Previous Figure 5.3 provides an example of routing, with examples of what each of the messages would look like with strict routing shown next. Provisional responses have been excluded for brevity. The examples are of a dialog between a UAC and a UAS, with proxies in between. The first leg of the call from the UAC to the first proxy would look like this:

```
INVITE  sip:raleigh.bellhead.com  sip/2.0
VIA: SIP/2.0/UDP  pchome101@aol.com:5060;branch=z9hG4bK74gh5
FROM: Deby Russell <sip:deby.russell@aol.com>;tag=9hz34567sl
TO: Travis Russell <sip:travis.russell@tcg.com>
MAX-FORWARDS: 70
CALL-ID: 82167534@126.18.27.0
CSeq: 10001 INVITE
CONTACT: Deby Russell <sip:deby@126.18.27.0>
CONTENT-TYPE: application/sdp
CONTENT LENGTH: 154
```

The first proxy then appends the message with its own address for recording the route.

```
INVITE  sip:losangeles.hometel.com  sip/2.0
VIA: SIP/2.0/UDP  raleigh.bellhead.com;branch=z9hG4bK32ad1
VIA: SIP/2.0/UDP  pchome101@aol.com:5060;branch=z9hG4bK74gh5
RECORD-ROUTE: <sip:raleigh.bellhead.com>
FROM: Deby Russell <sip:deby.russell@aol.com>;tag=9hz34567sl
TO: Travis Russell <sip:travis.russell@tcg.com>
MAX-FORWARDS: 69
CALL-ID: 82167534@126.18.27.0
CSeq: 10001 INVITE
CONTACT: Deby Russell <sip:deby@126.18.27.0>
CONTENT-TYPE: application/sdp
CONTENT LENGTH: 154
```

The first proxy then forwards the request on to the next hop in the network, where the second proxy address is added to the request.

```
INVITE  sip:travis.russell@tcg.com  sip/2.0
VIA: SIP/2.0/UDP  losangeles.hometel.com;branch=z9hG4bK42al3
VIA: SIP/2.0/UDP  raleigh.bellhead.com;branch=z9hG4bK32ad1
VIA: SIP/2.0/UDP  pchome101@aol.com:5060;branch=z9hG4bK74gh5
RECORD-ROUTE: <sip:losangeles.hometel.com>
RECORD-ROUTE: <sip:raleigh.bellhead.com>
FROM: Deby Russell <sip:deby.russell@aol.com>;tag=9hz34567sl
TO: Travis Russell <sip:travis.russell@tcg.com>
MAX-FORWARDS: 68
CALL-ID: 82167534@126.18.27.0
CSeq: 10001 INVITE
CONTACT: Deby Russell <sip:deby@126.18.27.0>
CONTENT-TYPE: application/sdp
CONTENT LENGTH: 154
```

Notice also that the *MAX-FORWARDS* header is decremented each time the message is forwarded by a proxy. Also note that the *TO* header still does not have the *TAG* field appended; therefore, a dialog has not been established.

Once the request reaches the UAS, a response is generated and returned back to the UAC. The response sequence might look something like the following examples:

```
SIP/2.0  200 OK
VIA: SIP/2.0/UDP  laptop107@tcg.com;branch=z9hG4bK32de5
FROM: Deby Russell <sip:deby.russell@aol.com>;tag=9hz34567sl
TO: Travis Russell <sip:travis.russell@tcg.com>;tag=1df789jkf
MAX-FORWARDS: 70
CALL-ID: 82167534@126.18.27.0
CSeq: 1001 INVITE
CONTACT: Travis Russell <sip:travis.russell@135.18.10.10>
```

The UAS in this message has appended the *TO* header with its own *TAG* value and has changed the *CONTACT* header to reflect the address of the recipient.

The first proxy in the reverse direction will then append the message with its address in the same fashion you saw in the forward direction.

```
SIP/2.0  200 OK
VIA: SIP/2.0/UDP  losangeles.hometel.com;branch=z9hG4bK32de5
VIA: SIP/2.0/UDP  laptop107@tcg.com;branch=z9hG4bK32de5
RECORD-ROUTE: <sip:losangeles.hometel.com>
FROM: Deby Russell <sip:deby.russell@aol.com>;tag=9hz34567sl
TO: Travis Russell <sip:travis.russell@tcg.com>;tag=1df789jkf
MAX-FORWARDS: 69
CALL-ID: 82167534@126.18.27.0
CSeq: 1001 INVITE
CONTACT: Travis Russell <sip:travis.russell@135.18.10.10>
```

The first proxy in the reverse direction then forwards the response to the second proxy in the reverse direction, which also adds its address in the *RECORD-ROUTE* header.

```
SIP/2.0  200 OK
VIA: SIP/2.0/UDP  raleigh.bellhead.com;branch=z9hG4bK32ad1
VIA: SIP/2.0/UDP  losangeles.hometel.com;branch=z9hG4bK32de5
VIA: SIP/2.0/UDP  laptop107@tcg.com;branch=z9hG4bK32de5
RECORD-ROUTE: <sip:raleigh.bellhead.com>
RECORD-ROUTE: <sip:losangeles.hometel.com>
FROM: Deby Russell <sip:deby.russell@aol.com>;tag=9hz34567sl
TO: Travis Russell <sip:travis.russell@tcg.com>;tag=1df789jkf
MAX-FORWARDS: 68
CALL-ID: 82167534@126.18.27.0
CSeq: 1001 INVITE
CONTACT: Travis Russell <sip:travis.russell@135.18.10.10>
```

The UAC now receives the responses from its request and processes the response. This includes the recording of the route used with each of the requests. This now becomes the route list for the *ACK* sent by the UAC as shown here:

```
ACK  sip:raleigh.bellhead.com  SIP/2.0
VIA: SIP/2.0/UDP  pchome101@aol.com:5060;branch=z9hG4bK713a2
ROUTE: <sip:raleigh.bellhead.com>
ROUTE: <sip:losangeles.hometel.com>
FROM: Deby Russell <sip:deby.russell@aol.com>;tag=9hz34567sl
TO: Travis Russell <sip:travis.russell@tcg.com>;tag=1df789jkf
MAX-FORWARDS: 70
CALL-ID: 82167534@126.18.27.0
CSeq: 1001 ACK
CONTACT: Deby Russell <sip:deby.russell@126.18.27.0>
```

The *ACK* is then sent through the network to the next proxy, which removes its address from the *ROUTE* header and forwards the message. It also places the address of the next hop into the request-URI.

```
ACK  sip:losangeles.hometel.com  SIP/2.0
VIA: SIP/2.0/UDP  raleigh.bellhead.com;branch=z9hG4bK32ad1
VIA: SIP/2.0/UDP  pchome101@aol.com:5060;branch=z9hG4bK713a2
```

```
ROUTE: <sip:losangeles.hometel.com>
FROM: Deby Russell <sip:deby.russell@aol.com>;tag=9hz34567sl
TO: Travis Russell <sip:travis.russell@tcg.com>;tag=1df789jkf
MAX-FORWARDS: 69
CALL-ID: 82167534@126.18.27.0
CSeq: 1001 ACK
CONTACT: Deby Russell <sip:deby.russell@126.18.27.0>
```

This continues until the *ACK* is received at the UAS. Once the *ACK* is received by the UAS, the session can begin. The final *ACK* in this example would look like this example:

```
ACK  sip:laptop107@tekelec.com  SIP/2.0
VIA: SIP/2.0/UDP  losangeles.hometel.com;branch=z9hG4bK32de5
VIA: SIP/2.0/UDP  raleigh.bellhead.com;branch=z9hG4bK32ad1
VIA: SIP/2.0/UDP  pchome101@aol.com:5060;branch=z9hG4bK713a2
FROM: Deby Russell <sip:deby.russell@aol.com>;tag=9hz34567sl
TO: Travis Russell <sip:travis.russell@tcg.com>;tag=1df789jkf
MAX-FORWARDS: 68
CALL-ID: 82167534@126.18.27.0
CSeq: 1001 ACK
CONTACT: Deby Russell <sip:deby.russell@126.18.27.0>
```

There are a number of reasons for using the *RECORD-ROUTE* and *ROUTE* headers (and therefore strict routing). There are many services that need to know when a session is released. For example, a presence server needs to know that a user has finished a phone call and is now available for another call.

The only way this is possible is if the proxies in the network are able to track each and every request and response for a session, and report via event notification to the presence server that a session has ended.

Strict routing also helps enable lawful intercept by routing all requests/responses for a session through the same proxies. This makes it easier to collect all call data for a particular session and correlate those messages for the creation of xDRs and reporting tools.

When routing a SIP request, the request-URI will identify the next hop in the network, employing the route list established for a particular subscriber. It should be noted here that the *TO* and *FROM* headers are never used for routing purposes. In fact, the network ignores these headers completely. They are only for the consumption by applications and user displays (such as calling name displays).

SIP Session Modification

Any session can be modified while the session is in progress. The media can be added, deleted, or changed, as can participants. A good example of this would be a Webinar, where participants come and go and the media changes from audio to video and file sharing.

When either UA wishes to modify a session in progress, it initiates a new *INVITE* using the same *CALL-ID*. The dialog ID is also the same which means the TAG values used to calculate the dialog ID are not modified from the initial request. The new *INVITE* contains the modifications and is referred to as a re-invite.

The re-invite will also contain all of the session description from the initial request. In other words, the entire session description including the modifications is sent to prevent confusion. This also allows stateful proxies to maintain an accurate status of the session.

The SDP will contain the actual changes, so the UAS receiving the re-invite will check the version header of the SDP (*v=*) to see if a change is being indicated, and will then look to see what was changed from the original request.

The UA that initiated the modifications then becomes the UAC for that portion of the session. Remember that only the UAC function can generate a request, but devices support both functions and can interchange roles while sessions are in progress.

There is one limitation for re-invites: They cannot be forked by forking proxies. The request-URI contains the address of the UA that the original session was established with, preventing forking of the re-invite.

If the receiving UAS chooses not to accept the modifications, then it can respond with 488 NOT ACCEPTABLE HERE without disrupting the initial session. The response only affects the modifications to the session. The *WARNING* header then provides an explanation as to why the modification is being rejected.

If the UAS should send a 200 OK but never receive an *ACK*, then it will respond with the *BYE* method. This will disrupt the entire session and result in tearing the session down. It is assumed that the dialog has been lost between the other entities and the UAS must therefore release itself and its resources from the session.

Since alerts (such as ringtones) are not typically needed for session modifications, the re-invite will not contain the *ALERT-INFO* header. Likewise, there is no need for the provisional response 180 RINGING (although 100 TRYING may still be used). This does not mean that user interaction is not supported, however.

For example, if video is being added to a call, the user will need to be notified by the application on his or her device that video is being added so that the user can select the video option. The application is responsible for notifying the user with whatever mechanism is appropriate.

If no user interaction is required, the *UPDATE* method can be used to modify a session. When using the *UPDATE* method, simple modifications can be supported without using the re-invite mechanism.

SIP Session Termination

Any UA can terminate a session. First, remember that a session is the result of a request receiving a positive response (1*xx* or 2*xx*). The completion of the handshake and the creation of a dialog must take place first. If there was any other response to the request, the session is then effectively terminated (actually, it never initiated).

The *BYE* method is used to terminate sessions. As seen in previous Figure 5.3, the *BYE* is sent in either direction, followed by the response 200 OK. This completes the process. The *BYE* method also terminates the dialog between the two devices.

This means that the *BYE* is not sent outside of a dialog. In other words, if there is no dialog or session established, then the *BYE* method cannot be used. The *CANCEL* method is used to cancel a request prior to a dialog/session being established.

When using the *CANCEL* method, the UAC initiates the message and sends it to the UAS. The UAS has not yet replied to the request; therefore, the UAS should have no knowledge of the request, which is why the UAC must always initiate the *CANCEL*. The UAS will then send the 487 REQUEST TERMINATED response to the *CANCEL*.

Should the UAC send the *CANCEL* and then receive a 2*xx* final response, it can send an *ACK* to establish the session, or it can terminate the session by sending the *BYE*.

The UA terminates a session for one of two reasons: the user wishes to terminate the session, or there was an error. An error could occur, for example, if a 200 OK is sent by the UAS but an *ACK* is never received. The UAS would then send the *BYE* message to terminate the dialog.

The *BYE* message must contain tags so that it can be correlated with a dialog ID and session. Otherwise, it will not be clear which session/dialog the *BYE* is terminating. The only response to the *BYE* is the 200 OK response used to confirm that the *BYE* was received and resources can be released.

6

Extending SIP to Support New Functions

In the early days of Voice over IP (VoIP), there were many interoperability issues. There were also many different protocols used to control the various sessions within a network, adding to the often chaotic implementation of an IP network. Many vendors supported only some of the protocols, electing not to develop to all of the standards that were being introduced for these networks. Instead they elected to carve their own paths and hope that they could win the influence of enough customers to make their platforms the new standard.

This was the biggest frustration shared by many traditional operators trying to implement VoIP in their stable, traditional telephony networks. These networks consist of tried and true technologies that are based on age-old standards with many years of maturity. Vendors have conformed to these standards, ensuring now that interoperability is a given. It wasn't always this way, though, and so it comes as little surprise that we are seeing the same interoperability issues today that we experienced when SS7 was first introduced.

Yet the attraction to VoIP for many is the ingenuity. There are many more unique and "killer" applications now available based on VoIP technologies that we probably would have never realized using the old "tried and true" technologies operating now. What is needed is a technology that can be standardized to a level that guarantees interoperability, while allowing for ingenuity and vendor uniqueness. That was accomplished through the introduction of the Session Initiation Protocol (SIP) and its support of extensions.

SIP was meant to be backward compatible with all earlier version of SIP to prevent operators' networks from becoming isolated due to non-compliance with some software upgrade. SIP was also meant to be interoperable with all operators' equipment in an effort to eliminate proprietary solutions in the network.

These were some of the issues experienced by many operators in the early days of SS7 and the IN. Some equipment did not interoperate with other equipment, forcing operators to purchase their infrastructure from one vendor (or one vendor's partners).

Of course this traps the operator and the subscriber as well, because they are both then limited to that vendor's solutions and accompanying features.

With SIP, operators can implement any platform that is compliant with RFC 3261 and be assured that it will be able to process SIP traffic. While there will still be proprietary SIP implementations, RFC 3261 dictates that all SIP entities must be compatible with one another by adhering to the requirements of the RFC. Other RFCs have been published to define the interactions with other SIP capabilities and functions such as messaging and other similar applications.

What this does is allow operators to implement a SIP variation supporting many unique functions and applications through customized SIP headers. These headers allow SIP to provide much more capability, while also ensuring that as SIP travels through other networks, the other networks are still able to recognize the headers needed to route the message to its destination without any degradation in service.

There have been many contributions from vendors developing SIP extensions. Some are published, and some remain proprietary. The 3GPP also realized that RFC 3261 did not address all of their immediate needs, and therefore they defined a number of extensions for use within their IMS architecture. Those extensions were documented in subsequent RFCs and published through the IETF. Those extensions are described in this chapter as well.

The Concept of SIP Extensions

By supporting extensions to the protocol, vendors are given greater freedom to create features and functionality that the standards bodies may not have thought of, while still ensuring interoperability in a mixed vendor network. This is also an advantage to network operators because it allows them to implement multivendor networks while still realizing the maximum in features and applications.

Just as important is the need to transport SIP messages through many different networks. This means that all networks must be able to understand and process a minimum set of SIP headers required for routing the message through the network.

The minimum set of requirements for SIP networks is defined in the RFC 3261. All vendors must conform to this minimum set of requirements and messages to ensure that the SIP messages that their equipment generates can be processed and routed through many networks end to end.

However, SIP introduces many more possibilities. SIP can be used to introduce many new features and applications that traditional networks cannot support today. SIP is a communications protocol that allows network entities to communicate instructions regarding any services and any media with one another.

Support for additional headers in the SIP protocol allows vendors to create their own unique applications and to support those applications through SIP signaling. Keep in mind that SIP does not actually enable anything beyond providing the communications between all of the servers and proxies that are responsible for actually delivering services.

By allowing new headers to be added at will by any vendor, the creators of SIP have provided a framework on which network operators and vendors can build and still be ensured that they are able to connect and communicate with any other network in the world.

This is the concept behind SIP extensions. SIP extensions can also be proprietary and known only to a single vendor. The extensions are then developed to support specific features and functionality that are developed only by that vendor, and therefore the extensions are not public knowledge.

The vendor can still route SIP messages in a standard SIP network, and any network element that receives a SIP message with headers it does not understand continues to process the SIP message, ignoring the headers it does not support.

How Extensions Are Documented

A vendor can also choose to have their extensions included in the standards by publishing an RFC for a specific application or function and submitting it to the Internet Engineering Task Force (IETF) for consideration.

The Internet Assigned Numbers Authority (IANA) also provides a valuable service to all those wishing to understand what extensions have been filed, offering a listing of all documented and ratified SIP methods, headers, and extensions.

The 3GPP has also submitted a fair number of standards identifying new applications and features based on SIP. The work of the 3GPP is very specific to implementations in a wireless environment, while TISPAN is developing standards relative to wireline networks implementing SIP. All of these organizations are focusing on a specific network architecture known as the IP Multimedia Subsystem (IMS), which is based on SIP.

There are also a number of RFCs that are published identifying how network elements should handle various extensions, or more accurately how a specific function or application works and what extensions have been identified to support the application. The IANA is a good source for finding these RFCs, at www.iana.org. Of course the IETF is always the best source for all RFCs, but it can be confusing to find the right document. The 3GPP also has numerous documents published for SIP implementations in traditional wireless and wireline networks. The IANA Web site will provide a listing that proves to be quite useful to anyone working in SIP.

Of course there are still proprietary extensions developed by vendors with no intention of ever publishing the extensions. These extensions are understood only by the vendors' equipment and therefore will not be supported by other vendors' systems. There is no need to document these extensions in a public forum. You will need to request the documentation from the respective vendors.

How Extensions Are Treated

When a SIP entity receives a SIP message with headers it cannot interpret, it ignores the portions of the SIP message it does not understand. Again the RFC defines a standard set of SIP methods (messages) that all SIP entities must be able to understand and process to ensure network operability as well as interoperability.

It is this minimum set of requirements that ensures that all networks can interconnect with one another and pass SIP messages forward. All networks must be able to support the basic functions of registration and routing according to RFC 3261.

If a SIP entity needs to know what methods another entity can support, it will send the *OPTIONS* method to the other entity. This allows other entities to discover what methods a network element can support prior to sending an invitation for a session. Hopefully this prevents sessions from being rejected because the network element received a message type (method) it did not understand.

The user agent that sends the *OPTIONS* message could also include the *REQUIRES* header to identify specific methods, codecs, extensions, or other support needed for sessions. The receiving UAS would then respond with a 2*xx*. The response would include the *ALLOW* header listing all of the message types that are allowed or supported by the UAS. The message body carries the Session Description Protocol (SDP), which would then identify the codecs and other resources supported by the responding UAS.

The *OPTIONS* method is only one method that can be used to determine what another network element can support. In the *INVITE* message itself the *REQUIRES* header can be used for the same purpose. The *ALLOW* header would be included in the 2*xx* response to the *INVITE*. The syntax and format of these messages is provided in greater detail in Chapter 2.

All of these mechanisms were introduced to allow network entities to communicate with one another their specific capabilities. This also allows operators to implement specific applications through specific entities, while limiting the support required by other entities to minimum SIP support.

Some Examples of Extensions

There are many examples of extensions and how they can be used in the network. These extensions are prefixed with the letter *P*, indicating they are "private" extensions. This can be somewhat misleading, though, since the intention is to publish these extensions so that equipment vendors can choose to support them in their respective network types.

For example, packet cable networks specify specific headers that need to be supported above and beyond those identified in RFC 3261. Any one wishing to operate in these networks should be able to support these extensions unless they plan on providing nothing more than basic SIP support.

Likewise, the 3GPP has defined a number of functions and applications that are required for the network architecture that they have introduced (the IMS). Any operator deploying an IMS network must conform to the 3GPP standards to ensure that all of the entities defined in the IMS architecture work properly and support the functions defined by the 3GPP.

Failure to support these extensions does not mean that equipment will not work in a SIP network. All network entities are required to support the functions and processes defined in RFC 3261. This includes all headers and methods and their respective parameters. If extensions are included in a message that a network entity does not understand, as we mentioned before, they are simply ignored and the SIP message is processed in terms of the portions of the message that the entity does understand.

What it does mean is that some equipment may not support additional functions and applications the operator needs according to the network topology being implemented, and according to the services the operator wishes to provide. The packet cable and IMS are good examples of networks that require the use of specific extensions to work as defined.

The following section defines extensions that have been ratified by the IETF and documented in the form of an RFC or 3GPP specification.

P-Access-Network-Info

This extension was introduced through the 3GPP, primarily for use within 3GPP (IMS) networks. The header provides additional information about the access technology that was used for accessing the network. For example, for a wireless subscriber using packet services, the header may look like:

```
INVITE sip:travis.russell@tcg.com SIP/2.0
Via: SIP/2.0/UDP homepc101@aol.com;branch=z9hG4bK2bndnvk
To: sip:deby.russell@aol.com
From: sip:russell@tcg.com;tag=346249
Call-ID: 2Q3817637684230998sdasdh10
CSeq: 1826 INVITE
Contact: <sip:russell@192.1.2.3>
P-Access-Network-Info: access-type= IEEE-802.11b
```

How this information gets used by the UAS is up to the service provider. One example might be providing specific services to the subscriber based on the type of access. If the subscriber is using a low-speed access link, for example, that subscriber may be restricted from using services that require higher-speed connections.

This can also be used for billing applications where the billing is dependent on the access type. If the subscriber uses an 802.11 network to access the service provider network, the 802.11 service may be provided through another operator, and there may be additional charges for using this access method.

Since this can be sensitive information, it is desirable to encrypt the information or delete it prior to transmitting any messages outside of the trusted domains. This is also implementation specific, so different operators may use this data differently.

Any proxies in the path of the message ignore the header, as this information is not used by intermediate proxies, but by proxies providing a service (such as the S-CSCF in the IMS). The intermediate proxies simply ignore the header and route the message forward; if the proxy is providing a service, however, then that proxy will interpret the header and delete it prior to forwarding to another network.

P-Answer-State

This extension was introduced to support push-to-talk applications in cellular networks. The way this would work in a SIP network could vary depending on the implementation. This header was introduced to provide a better experience for the subscriber using the application.

The intent is to minimize delay in the delivery of voice from the originator to the destination. What is different about this service and standard voice calls is that the devices themselves often have the capability to auto-answer an incoming call. This application works much like a walkie-talkie (think Nextel) and therefore needs a faster mechanism.

When using SIP, the destination device would have to return a 200 OK response prior to the voice being sent to the device. This would introduce a bit of a delay. To eliminate this delay, this extension was added. The extension is used by the application servers supporting the push-to-talk application itself.

These servers are capable of receiving the voice packets in advance of the destination device's actually responding to the initial request. The voice packets are buffered until the device has responded.

The servers supporting the destination devices will usually know if the device is configured to automatically answer an incoming push-to-talk call; therefore, when the servers receive an *INVITE*, the server will insert this header and forward the *INVITE* to the next hop. The header is inserted in the response back to the requestor, informing it that it is acceptable to send the voice packets prior to a final response. In this response the server includes this header indicating that it believes that the destination will be able to accept the invitation ("unconfirmed" header value) and is configured for auto-answer.

The server will then send this provisional response in the backward direction to notify the originator of the request that the message is being processed, and when the final push-to-talk server is reached, a 200 OK is returned by that server (where the buffering is to take place).

The originator can then begin speaking, with the packetized voice routed to the push-to-talk server. The server then buffers this packetized voice until the 200 OK response is received from the destination device, at which point the voice is forwarded directly to the device. The server buffering the voice does not forward the 200 OK response, as it has already sent this response to the UAC originating the session.

However, any servers in between the destination and the buffering server will see this 200 OK response with the *P-ANSWER-STATE* header having the value of "confirmed." There are provisions that allow the implementation of multiple servers supporting push-to-talk, with the server closest to the called party performing the buffering of the packetized voice. This is the server that is most likely to understand the configuration of the destination device, and whether or not it is capable of auto-answer (in which case it will automatically send a 200 OK response).

The contributors of this extension believe that sending the voice packets prior to receiving the final answer from the called party will expedite the delivery of voice to the device and provide a more positive and real-time experience for the called and calling parties.

P-Asserted-Identity

In SIP networks, the identity of the person originating a request can be questionable. This is because of the inherent ability within SIP to spoof an address, or to hide the *FROM* header contents altogether (as is the case with "anonymous" proxies). When this happens, calling party name display applications no longer work, and law enforcement agencies are unable to trace calls by identity. This also has an impact on the subscriber, who can no longer trust that the true identity of a caller is known.

This extension addresses these problems by adding a new header called the *P-ASSERTED-IDENTITY* header. This header contains a display name (used to display the name of the calling party on the display device) and the URI of the calling party. The intent is certainly that this is inserted by the network and therefore has been verified as the legitimate identity of the calling party (and not some spoofed address).

The header is added by proxies in the network within a trusted domain. If a proxy receives a request or a response that does not contain this header, the proxy first must authenticate the originator of the message to determine the true identity. After authentication, the proxy adds this header with the identity it obtained during authentication.

The proxy then forwards the message along with the header intact as long as the next hop is another trusted entity. If the message is to be routed outside of the network (into another network that is not part of the trusted domain), then the proxy removes this header prior to forwarding (if the subscriber has requested privacy). The *PRIVACY* header will identify whether or not the subscriber wishes to remain anonymous, and therefore the *P-ASSERTED-IDENTITY* header gets removed.

When a proxy receives a message from another proxy outside of its own trusted domain, and it contains this header, the proxy removes the header and performs authentication to verify the identity. It then inserts its own *P-ASSERTED-IDENTITY* header containing the identity it received during its own authentication.

A user agent never adds this header, and if it does, the header is not to be trusted by the network entities. The proxies will either reject the request or throw out the header prior to forwarding through the network, depending on how security is implemented. The UA can provide a "preferred" identity to be used in this header by inserting the *P-PREFERRED-IDENTITY* header prior to sending a message forward to the proxy. This header is described later. Here is an example of a *P-ASSERTED-IDENTITY* header:

```
P-ASSERTED-IDENTITY: "Travis Russell" <sip:travis@tcg.com>
```

P-Associated-URI

This header is used by the registrar to notify the UAC of other URIs that have been assigned by the network operator for use by the same subscriber. These are not registered URIs, but simply additional URIs that the operator knows about for the subscriber.

The header is included in the 200 OK response when the UAC is registering with the network using one of its identities. The additional identities are then returned by the registrar so that the UAC is aware of other identities that are authorized for this account.

This header is most likely used in networks where the identities authorized for any subscriber are set by the network operator instead of the users themselves. How these identities get associated with an account subscription will vary by operator, but chances are when a subscriber opens an account, that subscriber will be assigned telephone numbers (TEL URIs) and a set of public identities to choose from. These then become a part of his or her subscriber profile.

The use of this header is of course completely optional. However, if the network operator is going to support this extension, then the header must be inserted in all 200 OK responses to all *REGISTER* messages received, including re-invites. The header may contain more than one identity, or it may not contain any identities (be blank), but it must be present if it is going to be supported.

The UAC can then use any of the identities provided when sending other requests. The identities can be used in the *FROM* header. The UAC can check to see if any of the URIs are registered by sending a *REGISTER* message to the registrar. The UAC populates the *TO* header with its own address but does not send a *CONTACT* header.

The registrar in turn will send the 200 OK response with all of the registered URIs in the *CONTACT* header. If there are no addresses in the *CONTACT* header, then none of the addresses are registered.

P-Called-Party-ID

This header is inserted by the SIP proxy that is serving the destination of an *INVITE*. When an *INVITE* is sent to the destination, the request-URI will provide the address of record for the subscriber, as well as an address in the *CONTACT* header. The UAS or any SIP proxy handling the routing of the request may redirect responses directly to the *CONTACT* header address, bypassing the proxies entirely.

When this happens, the request-URI no longer provides the address that the request was originally directed to, causing the UAS to become confused as to what the original target address was. The serving proxy then uses the request-URI from the original request to populate this header.

This is sometimes necessary if a subscriber is using multiple identities. For example, the request may have been sent to a subscriber's business identity but redirected to the registered address. The proxy redirects the request to the registered address, losing the identity originally addressed. The UAS then does not know if the request was directed to the personal identity or the business identity for the subscriber.

This header then allows the UAS to handle the request properly, using the URIs that are contained in the registrar (remember, some of these identities can be administered by the service provider). The UAS can then apply unique alerts (ringing, for example) based on the identity originally targeted.

P-Charging-Function-Addresses

This header was also introduced through the 3GPP for use within IMS networks. In the IMS architecture, there is a whole new billing function defined. Several new entities have been introduced specifically for the purpose of billing within the IMS, and new procedures have been defined.

To further support billing within the IMS domain, some new headers were introduced. This is one of those headers. Its purpose is to provide the address of the billing function that is providing billing (or charging) information. Two types of functions are defined for use in the IMS architecture: the charging collection function (CCF) and the event charging function (ECF).

The CCF is used for postpaid billing applications. The address of this function is provided within this header. The ECF is used to support prepaid billing applications, and its address is also provided within this header.

These entities are distributed throughout the network, thus the need to be able to address the appropriate charging function. The actual address for each of the entities can be provided through numerous means, according to the operator's implementation. For example, the addresses may be provided through the S-CSCF within the IMS domain and inserted when a request is received.

Whenever a network entity wishes to provide billing information to the billing system, it will use the addresses provided in this header to route charging data using a separate network from SIP. DIAMETER is the protocol used within the IMS domain for this purpose. This header, then, is providing the address information for the charging entities to which this billing/charging data is to be sent.

Obviously charging data is very sensitive and should not be sent outside of the home network. For this reason, this header is never sent through other networks outside the home network. This and all other charging headers are deleted by the gateway proxies prior to sending messages outside of the network. Here is an example of this header:

```
P-Charging-Function-Addresses: ccf=193.3.2.1; ccf=193.1.2.3;
ecf=193.3.2.1; ecf=193.1.2.3
```

This example shows two charging entities: the ccf and the ecf at four different locations. The IP address is provided for each one. This address is then used to reach the charging entities using DIAMETER (which is also a packet protocol supporting IP transport).

P-Charging-Vector

The charging vector is used for the correlation of charging data. There are several parameters that are used to support this function: the IMS Charging Identity (ICID), the address of the SIP proxy that provides the charging data, and the Inter-Operator Identifiers (IOI).

All three of these identifiers collectively compose the charging vector. The *ICID* specifically is a unique identifier that is used for charging data correlation. The charging function uses this identifier to determine the session or dialog that the charging information is relevant to. The RFC suggests using a unique identifier created by combining the identifier and the IP address of the SIP proxy.

The IOI simply identifies the networks on both sides of the session. This would only be the case if the originating network were different from the terminating network. Otherwise, the network is the same and there is no need for *IOI*.

This header is used in conjunction with the *P-CHARGING-FUNCTION-ADDRESS* header, as well as the *P-ACCESS-NETWORK* header. Together these headers provide a fair amount of information that can be quite useful to SIP networks.

As with all charging data, this header is only used within a trusted domain and is not sent outside of the home network. The exception to this rule is if the outside network has an agreement in place with the home network, in which case it is considered a trusted domain.

Following is an example of this header:

```
P-Charging-Vector: icid-value=4321ab6789c; icid-generated-at=193.1.2.3;
orig-ioi=tcg.net; term-ioi=aol.net
```

P-Early Media

In SIP networks, the initiation of a session can sometimes result in conversation being chopped off, especially when the session first begins. This is due to the very nature of how a session is established, and when the medium is actually "cut-through." In the PSTN, the voice facility (trunk) is connected end to end in one direction to allow for tones (such as ringback) and digits (DTMF) to be transmitted. This is necessary for the sole purpose of supporting the call, not for the benefit of the user.

In a SIP network, the offer/answer model can result in a party answering his or her phone prior to any media being made available. For example, a perfect session would mean that the *INVITE* would carry the offer (in the SDP) for the session. The response, then, would provide the answer (again in the SDP), and the session would be established. The facility is connected and the voice is sent after the 2*xx* response is received by the UAC.

However, this is not always the case. Sometimes the offer is carried in the response itself, and the answer is provided in the *ACK*. This results in the voice being chopped because the UAC cannot provide resources for the session until the answer has been sent.

This header was introduced to solve this problem. There are many instances where media need to be delivered prior to the session being completely established (as in the PSTN model). The receipt of DTMF tones for an interactive voice response (IVR), for example, warrants early media being made available and connecting, as does an announcement being delivered to the calling party prior to the call actually being connected.

The purpose of this header is to notify originators of a request when early media need to be supported (this, again, is media transmission prior to the establishment of the session). Of course, care must be taken in SIP networks to prevent actual user data from being allowed during this time. The network proxies are responsible for ensuring that no user data is allowed to be sent during the early media period.

This is analogous with PSTN networks where the trunk is cut through in the backward direction for the sole purpose of signaling tones and announcements, as well as digit collection, prior to the called party answering.

P-Media-Authorization

It has been suggested that some calls will require a higher QoS than others. This requires some fashion of enforcing QoS policy for authorized calls. This header was added as a means of communicating an authorization key to the UAC wishing to have QoS enforced for a specific request.

The header is added by a proxy in the network that is configured to provide QoS support. The proxy must interface with an application providing the QoS authorization, so this does not necessarily mean that the proxy provides the authorization itself.

The proxy simply provides an authorization token. The token must then be sent by the UAC to the entity providing the resource. This is outside the scope of SIP and is supported by other protocols (such as the Resource Reservation Protocol [RSVP]). A media gateway, for example, may interface to another function providing a media reservation. As part of the media reservation, QoS is enforced by the media resource reservation function, ensuring that a resource will be made available to the UAC. The UAC provides the token provided in this header to this reservation function to show that it is authorized to receive QoS.

For example, a subscriber initiates a call by dialing digits. The UAC then inserts these digits into an *INVITE* and forwards the *INVITE* to the destination. The proxies within the network route the message to its destination, where the UAS examines the session description (SDP) to ensure it can provide the necessary resources for the session.

The UAS then returns a final response to the UAC agreeing to support the session with the required resources. The message is then forwarded back through the same path as the initial *INVITE*. The proxy responsible for supporting QoS will then examine the SDP portion and send this information to a resource reservation function.

The resource reservation function then verifies that the proper bandwidth is available to support the session, and provides an authorization token back to the proxy. This authorization token includes where the reservation is being held. The exchange of reservation information is exchanged using something other than SIP (such as RSVP).

The UAC then receives the response containing the *P-MEDIA-AUTHORIZATION* header with the authorization token. This is then used by the UAC to send to the reservation function using RSVP, for example. The media path is then secured with the requested QoS supported.

SIP is really providing a means in this case for communicating the proper credentials to be used by the reservation system, rather than providing the QoS function itself. This is important to understand, as SIP really is nothing more than a means to communicate call control information between the many different entities in the SIP network.

P-Preferred-Identity

This header is used by the UA to provide the proxy with a preferred identity it wishes for the proxy to use in the *P-ASSERTED-IDENTITY* header. The proxy does not have to accept this identity as a legitimate identity but can authenticate the user instead (thereby ignoring this identity and using the identity discovered during authentication instead).

The value provided must consist of a SIP URI, or a TEL URI. There can be multiple headers, but one must contain a SIP URI and the other must contain a TEL URI. Otherwise, only one header is allowed. This header is removed by the proxy prior to forwarding on through the network.

P-Profile-Key

The 3GPP introduced this extension for use in IMS networks. In an IMS network, a user database identifying all of the characteristics of a subscription such as permissions, identities, and location is found in the network core. Access to this user database(s) is only granted to a proxy referred to as the serving–call session control function (S-CSCF). The user database itself is referred to as the home subscriber server (HSS).

The HSS also contains identities for the various services being provided by the operator network. Services are hosted on application servers and therefore must be identified through the use of the same SIP addressing schemes as subscribers. This means the various services will also be identified by URIs.

These service identifiers have specific addresses so that the application delivering the service on an application server can be addressed. They also have "wildcarded" identities that are associated with them as well.

When a proxy is handling a request in the IMS, it may become necessary for the proxy to be able to access the user database and learn of the service identities associated with delivering the requested service(s) to a subscriber. Each proxy may need to perform this query, which places unnecessary burdens on the proxies and the user database. In the IMS, for example, a request may be received by the interrogating–call session control function (I-CSCF) requiring the I-CSCF to query the user database.

After forwarding the request to the S-CSCF, the S-CSCF would also have to perform this query, unless the service identity could be carried forward in the request as part of the SIP request. This requires a new header as well as knowledge of the proper key to access the user database.

This header contains the key needed for the proxies to be able to access the user database. No explanation is given as to how the key is acquired or how the header is used at this time, so there is little explanation as to how this works in an IMS environment.

P-User-Database

Like the *P-PROFILE-KEY* header, this header was introduced by the 3GPP for use in the IMS. The issue to be solved in the IMS is multiple proxies accessing the HSS function. The *P-PROFILE-KEY* header is used to communicate the access key for the database, while this header is used to provide the URI for the HSS serving the subscriber.

When a request is received by an I-CSCF, the I-CSCF must determine which HSS is serving that subscriber. To do this, the I-CSCF queries a subscriber location function (SLF). The SLF identifies the HSS serving a subscriber. Once the I-CSCF receives the address of the HSS serving the subscriber addressed in a received request, the I-CSCF forwards the request to the S-CSCF.

The S-CSCF must then also determine which HSS is serving the subscriber, and it will also query the SLF to find the HSS. This means that at least two different proxies have accessed the same information for one session.

To eliminate this need and to optimize the use of the SLF, the 3GPP has defined this header to be used to provide the S-CSCF the address of the HSS as learned from the SLF. The key for accessing the HSS is provided in the *P-PROFILE-KEY* header, which is also added to the request.

The I-CSCF usually has responsibility for adding these headers as it tries to locate a subscriber. The S-CSCF then does not need to query the SLF for the same information, as it has been provided by the I-CSCF.

This information is never sent outside of the network, as this would compromise the security of the network's resources (namely the HSS). For this reason, the header is removed from the request by the S-CSCF prior to routing to the destination. This means that the header is used only between the two proxies: the I-CSCF and the S-CSCF.

P-Visited-Network-ID Header

The 3GPP has defined this header for 3GPP (IMS) networks. When subscribers roam outside of their home networks and register from a visited network, the registration is sent back to the home network. This is because SIP procedures within the 3GPP IMS architecture require that subscribers always register with their home networks no matter where they are located.

The home network will then register the subscriber if the visited network has a roaming agreement with the subscriber's service provider. If there is no such agreement, then the subscriber's service provider has no means for collecting revenues from the service being provided to the subscriber while the subscriber is outside of the network.

Likewise, the visited network lacks the ability to charge back to the home network for roaming fees. However, if the visited network has a roaming agreement with the home network, it would then identify itself using this header. The home network is then able to determine if there is a proper agreement in place, and apply the appropriate charges back to the visited network.

Other networks that are transited may also elect to identify themselves using this same header; however, there is no requirement for this to take place. Other networks may elect to do nothing. In this case, the header will not be added and the home network will not be able to identify other transited networks.

The contents of this header consist of a simple text string identifying the carrier. The text string is unique for each operator, enabling easy identification without subsequent lookups in other databases.

The same information could probably be derived from the *VIA* header, but this would require the use of domain names in the *VIA* header. Since this is used for routing of the message through the network, this approach would require many queries to the DNS. This could overburden the DNS in busy networks. This is one reason for implementing this extension, enabling another vehicle for identifying visited networks without using the *VIA* headers.

In the event encryption is used between networks, there is an option to encrypt this header as well. This would make it impossible for the header to be read by any proxies other than the SIP proxy in the home network. For example, in the IMS the registrar function is provided by the Serving–Call Session Control Function (S-CSCF). The S-CSCF would be able to read the encrypted header because it knows the encryption keys.

Also remember that this header is not inserted by the UAC. Only a SIP proxy can insert this header as the message traverses the network. The operator may elect to have only its edge proxies (gateways, for example) serve this purpose. For example, in the IMS architecture, the Interrogating–Call Session Control Function (I-CSCF) should provide this header prior to forwarding the message outside its own network.

Packet Cable Extensions

The cable industry has added its own set of private extensions for use within their own packet cable networks. The cable industry has formed an alliance in the U.S., allowing all cable companies to interconnect with one another and bypass traditional telephone companies. They have formed a federation of registered and trusted companies that can then exchange data between their networks, treating one another as trusted domains.

The distributed call signaling (DCS) architecture defines new procedures for the exchange of billing data, as well as support for law enforcement requests. To support these procedures, the cable community has identified the following extensions for use within packet cable networks within the U.S.

P-DCS-Trace-Party-ID

This header is used within trusted domains to support call traces initiated either by the user or by the operator. The UA within the device adds this header within an *INVITE* to determine the identity of a calling party. The identity is then provided back to the user or to the operator back office functions for investigation.

If the UA is outside of the trusted domain, the UA will insert this header into an *INVITE* message. The request-URI is then directed to the address for the trace function within the network. The username portion of the request-URI is set to "call-trace." The network then provides the identity based on SIP messages (or other signaling) used to establish the call. Note that the *INVITE* sent by the requesting subscriber is not affiliated with the call that caused the trace to be initiated.

The *INVITE*, then, is a request to launch a trace. The network will then respond to the trace, usually by redirecting the session to a recording where the identity can be played back for the subscriber, or to some other address for processing (such as the operator's security department, for example).

P-DCS-OSPS

There are several situations in the PSTN where the operator provides a service from the operator services position (OSPS). This header supports three of those situations: busy line verification, emergency interrupt, and operator ringback.

Busy line verification allows callers who are receiving a busy tone when dialing a number to have the operator verify that there is a conversation on the line (instead of a phone off hook). The operator effectively breaks into the line to determine if there is a conversation on the line,and reports the status back to the requestor.

Emergency interrupt allows callers to request an operator to break into a conversation in progress and request that the parties release the line for an emergency incoming call. The operator can then connect the requesting party, or the requesting party can dial directly once the line is free.

The *operator ringback* is used to ring back to an operator for additional assistance, either manually or automatically as part of an application. All three of these services would result in the generation of an *INVITE* containing this header. The values for this header are *BLV*, *EI*, and *RING*.

The header is typically inserted into an *INVITE* sent over the OSPS trunk group by the media gateway controller (MGC) that is requesting the service. Only the *BLV* value can be sent in a new *INVITE*. Both *EI* and *RING* can only be sent in subsequent *INVITE*s related to a session in progress. The receiving UAS does not alert the parties involved in a session when the header is received with a value of *BLV*.

P-DCS-Billing-Info

This header is used to convey billing information provided by the proxies within a trusted domain. The header contains several parameters providing this data. The information is then used to provide billing systems with the necessary information for usage billing back to the billing system.

Several identifiers are provided to direct the billing data to specific entities for recording. The parameters provided are

- **Billing correlation ID (BCID)** This parameter consists of four bytes of time-stamp, an eight-byte network element identifier, eight bytes indicating the time zone, and a four-byte sequence number. The entire parameter is inserted into the header as a hexadecimal string.

- **Financial entity ID (FEID)** This is also a hexadecimal string identifying the financial entity to be used as the consumer of this data. The domain name is included after the "@" symbol to indicate the domain for this billing.

- **Record Keeping Server ID (RKSID)** This parameter identifies the servers to be used for recording the usage data being provided in the header.

The remaining parameters actually identify the usage itself and provide key account information. They all consist of E.164 numbers in the form of a TEL URI. These parameters are

- **ACCT-CHARGE-URI** This identifies the identity associated with the account to be billed for the usage recorded in the header. This is the equivalent to the "charged number."

- **ACCT-CALLING-URI** This parameter provides the identity of the calling party. It is different than the *ACCT-CHARGE-URI* in that billing could be to a different account. This identifies the actual calling party number for the call.

- **ACCT-CALLED-URI** This identifies the called party for the session and is the equivalent to the called party.

- **ACCT-ROUTING-URI** This parameter identifies the routing number discovered during an LNP query. It is applicable only to U.S. networks where Local Number Portability (LNP) is implemented. Other countries use different number portability mechanisms.

- **ACCT-LOCATION-ROUTING-URI** This parameter provides the local routing number (LRN) discovered during an LNP query. It is applicable only to U.S. networks where Local Number Portability (LNP) is implemented using LRNs for routing.

The UAC generating a request creates these headers, populates them with the correct information, and sends them as part of the initial *INVITE* request to the terminating UAS. There will obviously be proxies along the path that will process different aspects of this header.

For example, if the UAC is outside the trusted domain of the proxy, as would be the case for calls generated in another network, then the first proxy to receive the request would generate the header and all of its parameters, and forward the request to its destination.

If that destination is in another network, then the last proxy in the trusted domain will be responsible for deleting the header prior to forwarding the message to the destination. In other words, there can never be a case where billing information is sent outside the trusted domain.

The proxies send this billing information in the responses to a request as well, as long as the information stays within the trusted domain. The trusted domain, remember, can include other networks, as is the case with cable company federations.

The proxies will use another interface (non-SIP) for the exchange of billing information to an actual billing system. The purpose of this header is to exchange billing information specific to the session with other proxies that will be sending billing data to the same billing entities.

P-DCS-LAES

This header was introduced to support law enforcement requests for call trace and session content. Typically this is governed by local and federal laws requiring law enforcement to provide the operator with the appropriate subpoenas and warrants identifying the subscriber identity to be "tapped." Upon presentment of these documents, the operator must then provide the agency with the call content as specified.

In traditional PSTN networks this is accomplished through translations at the serving switch, which "forks" the transmission path to two different destinations.

The voice is sent to the address provided in the header. There are several parameters provided that are added by the proxy and used to direct content or call data to the lawful intercept function responsible for delivering this information to law enforcement.

The following parameters are included in this header:

- **LAES-SIG**   This parameter identifies the address for sending the call control data, such as called and calling party number and timestamp.

- **LAES-CONTENT**   Like the preceding parameter, this parameter identifies the address to send content such as voice.

Note that the UA that is typically performing this function is at the media gateway, since this is where the physical interface is for the transmission path. The MG, then, is responsible for inserting the correct information under the direction of the MGC. When LI is requested through administrative channels, the MG is configured to then intercept all sessions meeting the criteria of the court order. SIP is used to direct those intercepts to their proper destinations, but not to order the actual intercept.

P-DCS-Redirect

Redirect is also a function used to support law enforcement requests and is also managed at the MG. The redirect identifies the destination of the redirected session, usually an LI server providing content to law enforcement agencies.

The parameters provided in this header provide the called number (dialed digits), the redirect number (the number the session is being redirected to), and the number of times that redirect occurred.

As with all of these described headers and parameters, the contents are considered sensitive information and therefore are never sent outside of the trusted domain.

Security in a SIP Network

Security in the IP network is most certainly the most important part of any implementation for traditional service providers. This is because they have been operating in a highly secure network environment for decades. Today's legacy networks are very mature and highly secure networks, mostly due to the implementations by their operators.

Certainly if you ask operators of the world's leading telephone companies what keeps them up at night, you will quickly find network security at the top of that list. In fact it is security that has kept many companies from quickly embracing IP as a network transport and adopting IP for all their service delivery.

It is very ironic, then, that with security such a concern, many operators have such lackluster security measures. This includes established VoIP providers who have been victimized more than once by skillful and ingenious hackers and fraudsters. It is not that IP cannot be secured as much as the implementations are weak from a security perspective.

One of the reasons may be that securing a network can be difficult and requires a lot of specialized expertise. Most service providers do not want to accept the expense of hardening the security around their networks, and many more simply do not understand the measures that must be taken to prevent security breaches.

Should operators be concerned about security? According to statistics from both the GSM Association and from the Communications Fraud Control Association (CFCA), absolutely. Both of these organizations show that fraud is on the rise, mostly due to the deployment of IP technologies in the network.

In fact, fraud was on the decline until recently. As operators began deploying more and more VoIP networks, they began to see these networks come under attack by highly organized and well-funded organizations looking for ways to earn profits either by pumping their own traffic over other operators' networks, or by bypassing operators altogether and building their own peer-to-peer networks connecting through the operators' data networks.

These organizations are also highly educated and possess the resources necessary to breach even secure networks, although they would rather focus their efforts on those networks that are just too easy to breach and require little effort or resources.

When you look at the types of breaches being realized, you quickly learn that the attacks could have been prevented through some very simple measures. For example, a young man on the West Coast was sentenced just this year for hacking into thousands of computer systems looking for platforms that still had their default passwords intact.

He would then scan the systems looking for open ports that could be used for pumping his own traffic, or he would open up ports himself. Armed with the IP addresses of the systems he hacked and reconfigured, he worked with his partner (who actually perpetrated the whole business) to sell routes to legitimate telephone companies at cheap rates.

They routed millions of telephone calls through application servers sitting in businesses and universities across the U.S., charging telephone companies for access to their "network" and pocketing millions in profits. The main perpetrator has fled the country and cannot be found.

I bring this incident up as an example of how telephone companies (and in this case even well-established VoIP operators) can be defrauded of millions of dollars by even small operations. The hacker was later quoted as saying how easy it was to break into the computers and systems and change their configurations.

What was the most surprising aspect of this? That established service providers had deployed VoIP servers without changing the factory default passwords! The most rudimentary security measure had been skipped in these networks. In the many lectures I have done on the topic, I am always amazed at how many people do not change their passwords on a regular basis.

But passwords are only one measure. Changing passwords regularly is certainly one way of slowing down hackers and fraudsters, but there are many more measures that should be taken to further protect any SIP network.

Why is it then that we know security is important, and yet no one seems to be concerned about it (or at least not concerned enough to implement robust security measures)? Security can be very inconvenient not only for the operator, but for the subscriber as well.

Types of Network Attacks

There are many different forms of network breaches, but here we will talk about those that occur the most often. In fact, these are the types of attacks that are seen repeatedly not only in VoIP networks, but the Internet as well.

The Internet can serve as an excellent model of what could go wrong given its wide-open nature. Nothing can be trusted in today's Internet, and fraud and Denial of Service (DoS) attacks are rampant. Anyone who spends any time following security alerts and law enforcement activities in this area quickly understands the Internet is a risky place to do business unless you are prepared to lose a large portion of profits to fraudsters.

That is not to say that the Internet is a losing proposition; certainly there are many businesses thriving on the Internet. But they are also losing a lot of money from fraud, and they could be making a lot more if they didn't have to deal with security issues.

A case in point is the current concerns over Web sites and e-commerce. Certainly there have been numerous reports of sites being compromised, sensitive information being stolen from online sites, and consumers attacked with spam and phishing attacks on a regular basis.

Since SIP is a derivative of HTTP and SMTP, both Internet protocols that are exploited on a daily basis, it only makes sense to understand the vulnerabilities these protocols experience today to better understand the potential threats within a SIP implementation. The 3GPP and the IETF are also actively adding new procedures and enhancements to the protocol to further harden the protocol against network attacks.

Let's look at the various types of network issues, how they work, and how they impact the network operator and the subscriber. Then we will look at ways in a SIP network to prevent these attacks from happening (or at least some good measures to decrease the probability of their occurring).

Understanding how hackers approach a network is important to understanding how attacks begin, and what one should look for. Hackers approach attacks in stages, beginning with probing of the network.

Hackers first look for networks with easy access. This means a network with open gateways and platforms with default passwords still in use. They then begin looking for vulnerabilities in those platforms by scanning the platforms. Once a vulnerability is found, it is recorded for later use.

Many security breaches can be found at the scanning stage. In fact, you could use this approach on your personal computer to find possible viruses and bots that are scanning ports for open access (this is done through software utilities that provide information regarding the activities on your computer platform). This is the same technique used for network elements.

There are distinct characteristics to probing activities that you can look for to determine if there is pre-attack activity going on in the network—for example, multiple requests to the same server from the same UA, or maybe many requests from one UA to multiple servers and proxies.

Once a vulnerability has been identified and validated, it is used later for a larger attack. It is a combination of vulnerabilities that allows hackers to reach through many different networks, traverse through all of these networks unnoticed, and launch large-scale DoS attacks on unsuspecting networks.

Here are some common methods used when attacking a SIP-based network.

Registration Hijacking

SIP uses clear text messages, meaning anyone with a computer and some programming knowledge can "tap" into a network and capture SIP messages. This is different from bit-oriented protocols that simply transport "frames" of bits that when grouped into a defined format can be decoded to specific messages and parameters. ISDN and SS7 are good examples of bit-oriented protocols.

Since SIP uses clear text, if a hacker can capture these messages, that hacker is able to read subscribers' sensitive information such as their public and private identities. This information can then be used to "spoof" a subscriber. In other words, the hacker can use this information to gain access into the operator's network for his or her own use.

Let's say a subscriber has registered with the network, and the subscriber's location is recorded by the registrar. All calls and e-mails, instant messages, and any other session traffic are sent to this location.

The hacker then accesses the same network and uses the same subscriber information captured when "sniffing" or "tapping" the network to register with the network. Since the hacker is using the identity of the legitimate subscriber, her or she is granted access. However, the registration from the legitimate subscriber is not changed.

To the network it appears as if the subscriber has changed locations in the network and sent a new registration. The hacker has changed this through his or her own registration with a new location that now gets stored in the registrar. This means that all session traffic for the legitimate subscriber will now be sent to the hacker's destination instead of the subscriber's device.

Now eventually the subscriber will change the registration again while changing locations (provided the subscriber is mobile), but the hacker already has the subscriber's identities and network permissions and is therefore able to use this information to gain network access anytime he or she wishes.

Another means of accomplishing registration hijacking is to capture a subscriber's registration message and then "replay" the same message using a new location. This effectively registers the subscriber with the hacker's location. This is one of the more common means for hijacking registrations.

Session Hijacking

Session hijacking works much like registration hijacking, but this attack is used differently. A session hijacking is used to take over a session in progress. Session hijacking began with the World Wide Web.

A Web server is not a stateful server. A session consists of the UAS accessing a page from the Web server. If a subsequent page is accessed, that comprises another session (or at least new authentication from the user and the Web server). To alleviate the need of authenticating continuously when using a Web site, Web developers created the concept of cookies.

A *cookie* is nothing more than a data file, usually consisting of the session ID. The Web server sends the browser the cookie when the site is first accessed. The cookie is then sent by the browser application each time it accesses the Web server for another page to identify itself.

This concept was expanded for use with online shopping sites to maintain the shopping cart. When you visit an online site and wish to review your shopping cart, the cookie sends your authentication information so that you don't have to keep typing in your password.

If these cookies are intercepted and copied, they allow the interceptor full access to the session already in progress. This means the hacker now has access to all of your transactions and account information (so long as the session is still in progress). A cookie can be expired when the session is over, or it can be set for longer durations.

Many sites generate cookies using an algorithm that uses the timestamp and the IP address of the user to generate a unique identifier. This is an easy identifier for hackers to guess using random generators. Cookies themselves are somewhat controversial for many reasons, but mostly because they are misunderstood (many believe they are executables rather than data files). The usage of cookies is full of vulnerabilities, however, making them very susceptible to session hijacking, so their use within a SIP network is not highly recommended.

This is especially true if WiFi is going to be the access method into the network. WiFi networks are still very susceptible to eavesdropping, and the use of any clear-text transmissions is risky. Cookies can easily be captured through eavesdropping on access points, which would then compromise network services.

One simple check for hijacking is to check the time and date stamp of an incoming request or response. When a UAS receives a request, it should check the date and time with its own internal clock. If there is a discrepancy (more than 30 minutes, for example), then it is very likely that the request was intercepted and has been replayed with a changed destination address.

This happens when a session is hijacked and the message is captured for replay. The hackers will change the origination address of a message to their own and insert the message back into the network. If they do not update the *DATE* header, then the message will appear as if it has been traversing the network for some time, which is not normal—a session request that has been traversing the network for a long period of time (30 minutes is a long time) will most likely be deleted, as the interval specified in its *MAX-FORWARDS* header will have been exceeded.

Impersonating a Server

If you spend any amount of time on the Internet, you will most likely run into this form of security breach. There are many Web sites that look exactly like their official sites but are in fact hacker sites used for stealing information from unsuspecting consumers.

For example, there have been many documented cases where hackers have created sites to look like bank or credit card Web sites, using official logos and modeling the Web site to look exactly like the real Web site. When the subscriber types in the URL for these sites, they get redirected by a redirect server.

The redirect server is compromised by the hacker to send consumers to their Web site rather than the real Web site. Once the consumer has reached this site, they are asked for password information and other sensitive account information. This information is then stored on the server by the hacker and sold on the Internet to other hackers.

Sometimes the hacker will send out e-mails prompting consumers to go to the hacker's Web site to update their accounts or to claim refunds. Again once they reach these sites, they are asked for sensitive account information that can then be used to access the consumers' accounts. There are many breaches of this nature, from text messaging as well as e-mails.

Then of course there is always the possibility of compromising the DNS. This is known as DNS poisoning, where the DNS server is hacked and the IP address for specific servers is changed to a spoofed Web site or address. This is very likely in SIP networks where DNS is used to identify the IP address for domains and their applications.

The damage would be the same in a SIP network as anywhere else. For subscribers to be redirected to rogue application servers could have serious impacts on both the subscriber and the operator.

Tampering with Message Bodies

Since SIP is forwarded in clear text, it is not necessary for someone to have a decoder. Simply capturing the message is good enough, and this is easy to accomplish if one has access to any number of sniffing utilities and is connected to the same network. Once the hacker has captured a message, the message body and the headers in the SIP message can be modified.

For example, a hacker may capture an *INVITE* from a subscriber and change the *FROM* header to reflect his or her own address. This would provide the hacker access to a network he or she is not authorized to use, and would allow him or her to initiate sessions with other subscribers while pretending to be someone else.

There are other concerns with message tampering. Since text messaging is sent in a SIP message body and is also in clear text, a hacker could easily intercept SMS messages and change their content. This could be especially troublesome if messages were being sent by government agencies during catastrophic events.

Message tampering can be easily prevented, as can many of the scenarios we talk about in this section through encryption. Encryption prevents the hacker from being able to decipher the text, and therefore the hacker would be unable to change it and route it back to the network.

Tearing Down Sessions

Tearing down sessions is a concern but not as troublesome as a denial of service attack. Hackers could intercept requests from various subscribers and simply send a *BYE* message as a response (as if it came from a proxy or other network element). This would then terminate the session and cause it to be torn down.

This is more of a nuisance attack and is most likely to be launched against individuals, since it is far easier to launch a DoS attack against the network or network elements, with much more devastating and far-reaching effects.

Denial of Service and Amplification

Denial of service (DoS) attacks can take many different forms and can be launched using many different techniques. The easiest form is simply flooding the network with specific traffic types. For example, using a call generator, a hacker can send millions of *INVITE*s into the network attempting to flood the network with call requests.

We see these types of attacks take place many times, and they have even occurred in the PSTN. The use of SIP and IP provides much easier access to hackers, enabling these types of attacks much more frequently.

Another form of DoS attack involves application servers. By launching a flood of requests to an application server, the network element is immediately flooded and congested, taking it out of service. This can also happen with the DNS through flooding with DNS queries (similar attacks of this nature have occurred many times already). When the DNS is attacked, the entire network can be impacted, depending on where the server sits within the DNS hierarchy and whether or not redundancy has been implemented.

When redirect servers are used, the potential for amplification occurs. One message is forked into many different messages, which will also result in many different responses. An attack relying on amplification will send many requests toward targets known to be redirected, knowing that those targets will result in the request being multiplied.

As the request is multiplied, it is sent to several different destinations. Each one of those destinations will then send an appropriate response back to the forking proxy. It is not necessary for the responses to be successful responses, since the object is to create a large volume of SIP traffic in the network with the hope that this causes enough congestion to result in a DoS.

Since congestion is the ultimate goal, one request is obviously not enough. Nor is one target sufficient for such an attack. Therefore the attacker will use many targets and millions of requests, and will continue to send these requests until the congestion occurs.

The registrar can also be the target of such an attack. A hacker can register a subscriber listing many different user identities for the same subscription. This then provides the registrar with a list of multiple destinations for a request. The hacker then launches requests toward the public identity, which the registrar and proxies then send to multiple destinations based on the registration made by the hacker.

This is also considered amplification, as the registrar is "amplifying" the effects of the attack by sending to multiple destinations. The impacts are the same as a forking proxy, bearing the same results.

A similar kind of attack toward the registrar involves registering many different identities. Each identity consumes memory within the registrar, and therefore if a large number of registrations take place, the registrar runs out of memory.

This only works in open networks with little to no security where anyone can register and use the services of the network to route requests. Hopefully most networks will prevent this from happening just through simple authentication, preventing unauthorized subscribers from accessing the registrar.

The forking proxy in turn will continue to fork the requests into many requests and will continue to manage many responses from the various targets until it becomes congested (or some upstream network element becomes congested). At any rate, the end result is that the network becomes too congested to handle any further traffic and begins denying service to other subscribers. Some of the network elements may even crash due to the levels of traffic.

Bots and DDoS Attacks

Bots are simple scripts that are carried to a subscriber's device through Web sites, or other viruses transported using text messaging or e-mail. Even Bluetooth is highly susceptible to viruses, and cell phones have fallen victim to these as well.

A bot sits on a device and first looks for any open ports or connections that it can use to access the Internet. The bot then looks for other computers or devices connected to the same network and begins exploiting these systems. This is what makes bots especially dangerous, since they have the ability to self-propagate whenever the device is connected to a network. Even firewalls cannot prevent bots from infecting other computers when an infected computer connects behind the firewall.

But the most troublesome aspect of a bot is the ability to control the script from a remote server. Bots use the Internet Relay Chat (IRC) protocol to communicate with a URL programmed into the script. They then receive commands from a server connected to the URL that basically launches the script. The server is the command and control server.

Many bots are used for accessing Web sites and specific links on those Web sites to increase the number of "hits" to a Web site fraudulently. There are some businesses that make money based on the number of hits to a Web site or Web site link, and therefore the bots inflate the revenues at will.

Bots have now moved on to more menacing and threatening uses, causing DoS attacks in many networks. When one hacker successfully launches bots, they create their own little network of bots known as *botnets*. All of the scripts are now under the control of one person, which if launched against a target could be devastating for that target.

For example, if the bots were instructed to access the same URL at the same time, millions of machines could be accessing the same Web site simultaneously. This would certainly cause the server to crash because it would not be able to handle such a huge demand. The Web site would then be out of service.

This has happened more than once, causing businesses to close their Web sites for days. The losses quickly ramp up when your Web site, your sole source of revenue, has been attacked and shut down. Imagine now if that Web site was a bank, or a credit card company.

The largest botnet to date was recorded in 2007. Millions of computers were under the command of one botnet. If that botnet were directed to any URL, it would have a devastating effect. Of course, bots affect more than just computers.

Imagine if cell phones were infected with these bots. Already we are seeing cell phones become infected with experimental viruses, propagating through text messages and Bluetooth. If a botnet were to be amassed, the hacker would have millions of cell phones at his or her disposal. Imagine now if all of those cell phones placed a call at the same time to the same destination. That portion of the network would be out of service for an undetermined amount of time.

Even if the bots were instructed to do something as simple as send a text message to a cell phone, the end result would be a lengthy denial of service for that portion of the network. It has already been demonstrated that just a few text messages could be sent to a single cell site resulting in a lengthy outage. Millions of phones sending messages to many phones in a sustained attack could easily cause an entire network to quickly crash.

The effects of such an attack would be crippling. Not only would the operator lose substantial revenues, they would also lose the faith of the subscribers. Of course, if the attack were held in conjunction with a physical attack, the results would be especially devastating.

Fortunately law enforcement agencies have been diligently searching for botnets and have successfully stopped many of them, but they continue to proliferate. It has become more important than ever to prevent unauthorized attacks on your networks and protect the resources from being compromised.

Security Measures

Security remains one of the biggest issues in packet communications today. When one looks at the many security breaches and network attacks to date, security was seriously lacking. We already talked about one specific case where the perpetrator was able to hack into thousands of computers where even the simplest of security measures (changing the password) had not been practiced.

Security can be very robust and sophisticated, and it can be very simple. Don't be fooled into thinking that highly complex and sophisticated security measures will prevent attacks on your network. Some of the world's most secure networks have been breached.

However, this is no excuse for not implementing any security at all. When you look at the makeup of network attacks, there are basically two types of attackers: one that is looking for the easiest networks to breach (little to no security) and one who seeks out those with very high security (the challenge). Why, you might ask?

Think about those networks with very high security. There is usually something there worth stealing; otherwise, there wouldn't be so much security. If there is something of value to the hacker, then the hacker will attempt to break in. There is also the concept of a challenge. There are many who just want to be able to say they were able to do it as a personal challenge.

So what does this mean for the average network operator? This means you certainly have something worth stealing, so you definitely need to protect it. But you need not spend a fortune doing so, since we already know that the best security available is still breachable. You should take some basic measures at the very least to prevent being attacked just because your network is so easy.

Know that the more security you put into place, the least likely a hacker is going to choose your network unless there is something specific they want from your network. It's the same concept used in physical security.

For example, when putting an alarm system in a home or business, simply placing signs and stickers at every entry point identifying your home as alarmed is usually enough to keep most burglars away. Why? They know your home is going to be more difficult (not impossible) and they can find another house next door that will be easier to break into. This is true as long as there is nothing inside they are looking for specifically (again, if you have something they want, they are going to try and get it).

Thieves look for the easiest way that takes the least effort, so keep this in mind when securing the network. Making it difficult for the thief but somewhat non-intrusive to the subscriber is the difficult balance. There should be as much transparent security as possible so that the users of your services are not impacted (there is nothing worse than having to enter a password three or four times to access a service).

There are many ways to implement security. Many operators have implemented session border controllers as firewalls to their network. While this is probably a good practice, thinking this will stop network attacks is dangerous. Many recent reports have proven that the majority of attacks come from within the network itself, as well as outside the network. This means that robust internal security must be implemented as well as border security to prevent the network from being compromised.

Also bear in mind that attackers are now well-funded and highly educated individuals with the backing of highly sophisticated organizations. Organized crime and terrorist cells are the most common attackers today, and it then becomes necessary to build a security plan with this in mind. We are no longer fighting the teenager with nothing to do but hack computers. We are now talking about older individuals who pursue this for a living and do nothing more than break into networks for pay.

The most basic concept in network security is maintaining a trusted domain. This means that all interconnections are made with entities that are known and can be trusted to send legitimate traffic. In security terms, the trusted domain is often referred to as the "realm."

A realm can be considered analogous to the trusted domain. It is identified in security parameters within SIP using the domain name of the trusted domain. Everything within the realm is known by the network, meaning it knows the subscribers within its own control and has authenticated all users.

Anything outside of the realm is not to be trusted and should be authenticated prior to authorizing services. This is similar to the concept used in GSM networks, where a cell phone device must first register with the network and exchange credentials. Without proper credentials, the device is denied access to the network.

This example shows how the realm is identified within a security context in a SIP *INVITE*:

```
INVITE  sip:travis.russell@tekelec.com  SIP/2.0
AUTHORIZATION: Digest realm="tekelec.com"
```

When a subscriber is in their home network, and are establishing sessions within their home domain, they are still authenticated by their home network. However, the home network has all of the subscriber's credentials and can easily authenticate the subscriber without issue.

When the same subscriber is roaming in another network, the visited network does not have the credentials to authenticate the subscriber. Therefore, an agreement must be established between the home network and the visited network. This agreement allows for both network operators to exchange subscriber details necessary to authorize services, and to bill one another for services rendered when each operator's subscribers are roaming in the other operator's network.

The reason many phones do not work in other networks when you are roaming is quite simple. The operators do not have an agreement between themselves, and therefore your device is not granted authorization to access any services. This situation is getting a little better as operators partner with more and more networks, allowing their subscribers to roam in many different countries, but there is still a long way to go.

Take the cable industry, for example. The cable industry has established a federation among all of the cable operators. All members of the federation are considered as trusted domains and can exchange subscriber and billing information with one another. This means that the cable operators have effectively created a massive network of networks where their subscribers can get services as if they were in their home network without restriction.

Of course, cable operators have not gone wireless yet. Many have offered wireless services through established wireless carriers but have not launched their own networks. This is all rapidly changing as the cable industry quietly adds more and more services to its portfolio. When cable companies do build out their own wireless networks, they will be part of a huge network composed of many different trusted domains.

There are basically six aspects to securing a SIP network:

- Authentication
- Authorization
- Confidentiality
- Integrity
- Privacy
- Non-repudiation

Authentication requires the use of passwords and the exchange of credentials. We talked a little about this already. Whenever a subscriber registers his or her location with the network, the registrar should always challenge the initial registration. This challenge is explained in more detail in the course of this chapter.

It is unfortunate that many networks simply do not challenge registrations. Instead they verify the user identity and trust that the identity is true. This is one of the reasons there are so many attacks on SIP networks today. Simply challenging the registrations could eliminate many security breaches.

Authorization requires querying a database containing the basic account information for a subscriber. This account information provides the public as well as private identities for the subscription, and all the services the subscriber is authorized to access. This can be part of the authentication process to be most effective.

Confidentiality protects the subscriber and the subscriber's identity. It ensures that conversations cannot be snooped on, and that the subscriber can exchange information freely without the information being captured by someone else. This remains one of the big challenges for network operators, especially given the many tactics being used today to capture sensitive data from subscribers.

At the same time, it is equally important that the integrity of any data sent by a subscriber be sent intact without alteration. This includes any Web sites that may have been accessed as well. It is far too easy for hackers to access SIP messages and change the contents in an effort to change the service and where it is being delivered. It is also very easy to capture a SIP message containing text and alter the text message before it is delivered to its final destination.

Privacy can sometimes be an issue when it is openly provided to anyone. Today on the Internet there are anonymous services where e-mails and other messages can be directed in an effort to hide the address of the originator, and make it appear that the message came from someplace else. At the same time, privacy can also be offered as a feature to some clients who have a need for such a feature.

One example of this is law enforcement agents. Today when they make a call, the call does not give away their identity or the number they are calling from. This is an important service to law enforcement agencies and government alike, and it should be maintained even in the SIP domain.

Finally, non-repudiation prevents subscribers from accessing services and later denying they used those services. If the operator implements the right tools and audit systems, you should have total visibility to every network transaction that takes place. This includes any downloads that the subscriber may have made.

Today this is not the case, and the reason many operators are losing money on music and ringtone downloads. While it is true they are making money on these services, it is also a documented fact that they are losing even more money. One industry analyst firm reports error rates in the reporting of downloads to be as high as 80 percent. The reason for this is simple: no monitoring systems are capable of supporting all of the protocols used within a SIP network.

We have listed the various aspects of a security deployment. Now let's talk about the specifics behind security implementations in a SIP network. These are simple suggestions and recommendations and by no means an exhaustive description of all that can be done. Nor is this meant to be a very detailed, highly technical discussion on how these security mechanisms work. Always refer to the specific RFCs for the specifications.

Password and Access Controls

I put passwords first because this is the easiest and the most overlooked security measure there is. Passwords are painful for everyone, the consumer included. No one likes to have to deal with passwords every time they access their phone or every time they try to make a call.

What's ironic is that while we hate passwords, we would never think of leaving the house without at least locking the door, and many of us go one step further and set an alarm. Yet isn't it funny that typing in a simple password is too much of a bother?

The trouble with passwords is managing the passwords so that they cannot be compromised. This means implementing password aging, which would require everyone to change their passwords periodically. Some change their passwords every month, while others change passwords every three months.

There are several levels of password control within a network. Certainly the devices accessing the network should be protected themselves, but the network is the most critical asset to the operator, and every entity within the network should be protected as well.

This may seem obvious, yet most network operators do not change the passwords on their voicemail platforms, their gateways, or even their routers. Instead, they deploy these entities using the factory default passwords sent by the vendor. These passwords are well documented and advertised all over the Internet, which means the hackers know your passwords as well.

So the first, most rudimentary step to security in a network is to ensure that all entities within the network have their passwords changed from the factory defaults. But don't stop there, because passwords can be determined through programs developed just for this purpose. The passwords should be changed as often as can be tolerated.

This then requires an entire password management initiative. Passwords should be as long as possible (the password length makes it more difficult to crack) and incorporate as many character types as possible (numbers, letters, and symbols). This extends out to the terminals that are used to access and configure the network elements.

One of the best methods of password control (at least for terminal access) is the use of token IDs. A token ID is a completely randomized password that changes every 30 seconds or so. The password is based on an algorithm that only the software application (which resides on the host) and the password generator know. The password generator (my term) is really a little key fob–like device with a display of numbers that change every 30 seconds or more.

The numbers displayed represent the newest password, and when combined with a user's access password, they become a very difficult mechanism to hack. I strongly recommend the use of these for any terminal access, and for use on any PCs used to access any network element. A compromised PC that later connects to a network element could put the entire network at risk, given the nature of many of today's bots and viruses.

Encryption

Encryption solves a lot of problems. The best means of ensuring privacy while preventing message tampering is to encrypt the SIP message prior to sending it through the network. This does present some problems, though.

If forwarding a SIP message that is encrypted, the various routers and proxies in the network will not be able to read the addresses contained within the various headers. Therefore the entire SIP message cannot be encrypted, unless every network element is going to be provisioned with the proper decryption keys (not a very likely scenario).

The other issue is forwarding the SIP message outside of the trusted domain. Not everything can be encrypted, as this would make it impossible for the other networks to read the addresses necessary for routing. This is why there are headers that are not encrypted, while the remaining SIP message is encrypted.

Encryption does require all receiving entities to have the encryption keys so that they will be able to remove the encryption (decrypt) and return the message to its original state. Without the encryption keys, the network elements cannot do anything with the message, so care must be given in deciding how to encrypt and when to encrypt. There are numerous ways to implement encryption.

One approach is to encrypt only messages leaving the network. This means there are no encrypted messages internally, since all of the elements are within a trusted domain. The problem with this approach is that most attacks take place from within the network. Therefore encryption for outbound SIP messages does not solve anything. There should be encryption internally as well as externally.

When a message is encrypted, some headers will remain in "clear text," which means they are not encrypted. The headers that are encrypted include:

- Subject
- Accept
- Alert-info
- Expires
- Supported
- User-agent
- Reply-to
- Accept-encoding
- Error-info
- In-reply-to
- Unsupported

- Server
- Organization
- Accept-language
- Authentication-info
- Require
- Retry-after
- Warning

The message body itself is also encrypted, which could include the Session Description Protocol. Since the SDP is encrypted, any proxies that need to read the SDP portion of the message will need to be able to decrypt the message. This is another consideration when implementing encryption, since these elements will have to know the encryption keys. This may include firewall proxies as well.

And don't forget the network receiving the SIP message. The network elements receiving the SIP message will also have to know the encryption keys in order to process the SIP requests. This is another reason that network agreements have to be made to ensure that the interconnecting networks can be trusted.

Transited networks do not have to know the decryption keys, as they do not need to know anything more than where to route the message. The headers used for routing will contain enough information for the SIP message to reach its final destination. Transport Layer Security (TLS) is a good means of providing encryption between networks, while SIP Secure (SIPS) is designed for use within a trusted domain. TLS works at the TCP layer and is best when used between two networks where two network elements do not know each other. TLS cannot be used end to end.

SIPS is used in the request-URI to indicate that TLS should be used to transport the request/response to the designated domain. Once the domain is reached, local policy determines the treatment to be used within that domain. TLS can also be used within a network, although it is best suited for interconnections with other networks.

RFC 3261 specifies that TLS is to be used at proxies, redirect proxies, and registrars when interconnecting with other networks. They should also possess a site certificate for authentication. These proxies also must have the ability to validate certificates from other trusted sites, by storing the certificates from these sites (usually elements from within its own trusted domain).

IPsec is best within a network between trusted elements. IPsec works within an "enclave" or trusted network, implemented by each of the network elements at the operating system level. Security gateways can also be used to create virtual private networks (VPNs) to make the network more robust.

Another approach to encryption is tunneling a SIP message within another SIP message. The original SIP message is encrypted and then encapsulated within another SIP message for routing. The routing information from the original SIP message is used to populate the routing headers in the outside message, but nothing else is given in the outside message.

A proxy then will only read the request-URI, *VIA, RECORD-ROUTE, ROUTE, MAX-FORWARDS*, and *PROXY-AUTHORIZATION*. These are the only headers that can be modified as the message is routed through the network. When the UAS receives a message that has been tunneled in this fashion, it compares the encrypted tunneled message with the outer message. If there are any other changes to the contents, the message is considered compromised and the original message is rejected.

To allow the sender to remain anonymous, the *FROM* header in the encapsulated message can remain as is, but the outer message *FROM* header can be set to anonymous. This helps prevent the identity of the sender from being detected while the message is transiting other networks while still providing the identity once the message is received. Of course, this header should never be trusted to begin with, since it is too easily spoofed. The network should always rely and trust only the *ASSERTED IDENTITY* header.

Tunneling messages prevents messages from being hijacked, modified, and then replayed into the network. As the original message has been encapsulated and sent within an S/MIME body, it should not have been altered. When the UAS compares the encapsulated message with the outer message, it will identify whether or not there have been any other alterations to the original message. Both the encapsulated message and the outer message are duplicated.

If any of the headers in the outer message are different than the headers within the encapsulated message, the encapsulated message of course takes precedence. The headers in the outer message are discarded.

Everything we have talked about so far, encryption and tunneling, applies only to the SIP messaging and not the bearer traffic itself. The bearer traffic should also be encrypted to prevent unauthorized access. Obviously the bearer traffic contains much more value to the hacker than just the SIP signaling, so it should be adequately protected as well.

The exception to this is in the case of the *MESSAGE* method. Text messaging uses the *MESSAGE* method to deliver the "bearer traffic," which in this case is a text message. The text message itself is carried in clear text within the message body of the SIP request.

Since the text is in clear text, it becomes even more important to ensure that the message body is encrypted. S/MIME should also be used within the body text to further protect the message when it is transmitted across multiple networks.

The SIP protocol does support communicating between network elements regarding the security mechanism to be used. The security mechanism is encryption, and the communications between entities are necessary to communicate the encryption schemes supported by the various nodes.

There are three headers defined as extensions specifically to support the SIP security mechanism; *SECURITY-CLIENT, SECURITY-SERVER*, and *SECURITY-VERIFY*. These are used to communicate to either the UAC, UAS, or upstream proxies the method to be used.

A UAC can query an upstream node to determine the encryption methods it supports prior to sending a request. It does this by sending the *OPTIONS* header. The receiving entity then returns a response containing a list of methods it supports. If the responding entity is the UAS, it would return the *SECURITY-SERVER* header contained in its response.

The *SECURITY-SERVER* header would then list the methods supported. Possible values include *TLS, DIGEST, IPSEC-IKE,* and *IPSEC-MAN*. The receiving node then returns a request along with the *SECURITY-VERIFY* header containing the relevant security keys. A UAC uses the *SECURITY-CLIENT* header to send a list of support encryption methods to upstream nodes.

Authentication and Authorization

One method of authentication is through the use of certificates. This is implemented in many Web sites today. When you set your browser to check for certificates, and you access a network resource (such as a Microsoft Web site to purchase software), your browser will ask the Web site for its certificate.

These are public certificates that are advertised to "trusted" communities so that they are able to access the sites. This works by providing the browsers with the key for the "trusted" Web site. The key is used to determine if the certificate is valid or not. Only the key holder and the application accessing the site know the key.

The key can also be provided when a subscriber signs up for a new service, such as online banking. The online banking institution would exchange the key with the subscriber upon signup for the service. The subscriber and the online banking service then are part of a trusted domain.

The certificate is associated with a specific URI using a cryptic identity. Only the receivers with the proper key can decrypt the identity and therefore authenticate the site. This concept is not perfect by any means, but it is one way of at least making it more difficult to steal services from networks.

Anytime a service is being provided, and a subscriber sends a request that is of questionable origin (has not been authenticated before), the application server should reply with a 401 Unauthorized response, forcing the UAC to send the proper credentials prior to accessing the service.

When the UAC receives this response, it should resend the *INVITE* again but include the *AUTHENTICATION* header in the message. The *CALL-ID* should be the same as the original *INVITE* so that the server knows that this is a second attempt to access services and it is in response to a previous challenge.

The *AUTHENTICATION* header provides the application server with the necessary credentials so that the server can provide the subscriber with the requested services. This should be practiced in all networks to prevent unauthorized access to services.

Authentication can be applied to application servers as well. As when using a certificate, each application server would be configured with a unique encryption key. Only authorized UAs would receive the proper key. When they receive a request from an application server, they would then be able to challenge the application server (the same process used against the UAC in reverse).

This would help prevent rogue application servers from sending requests to unsuspecting subscribers, and it would allow devices to challenge the servers they access. Since it is encrypted, it should be more secure than certificates.

When the *MESSAGE* method is used for forwarding text messaging, application server and proxy authentication should be enforced to prevent spoofing and spamming. As mentioned in the earlier section "Encryption," the *MESSAGE* method is used to deliver a text message. The text message is delivered in clear text within the SIP message body. This makes it vulnerable to attacks unless encryption and authentication are enforced.

A *DATE* header should also be enforced to ensure that there is a timestamp for every message. Any time a *MESSAGE* is received with a timestamp indicating it is more than several minutes old, the message should be rejected by sending the response 400 INCORRECT DATA OR TIME.

The exception is when a store and forward function is used to deliver these messages. When a store and forward server is being used, the message will of course always be more than a few minutes old, since it was delivered to be stored and forwarded when the subscriber became available.

Strict Routing

This is a concept that the 3GPP introduced to be used within IMS networks. In a SIP network, a proxy is able to use any available route. The routing decisions are made in real time by each of the routers in the path, based on current availability and traffic conditions. This means that a request may take one path, while its responses may take another path. This, of course, makes it impossible to trace any of the messages or to capture all of the messages related to one session without capturing everything in the network and using a correlation engine to associate each of the requests/responses.

The exception to this rule is when stateful proxies are used, since a stateful proxy must see every response for a request in order for the proxy to be able to monitor the state of the session. If the proxy does not see all of the responses, it will not be able to determine if the request was successful, and it will not be able to determine whether or not a session was terminated (unless it sees the relevant messages).

Because of this type of deterministic routing, hackers are able to use man-in-the-middle attacks to intercept sessions and have them rerouted to another address. They are able to use replay to send copied messages back into the network without detection, even if they suddenly route responses to a different address than what was registered.

With strict routing, the routing path is recorded as the subscriber registers in the network. The *REGISTER* message contains the *RECORD-ROUTE* headers, used to collect the addresses of each of the proxies in the path. These addresses are saved in the order they were added, so that a route list can be established for the subscriber.

The route list then becomes part of the subscriber's registration. Stateful proxies in the path then also store the route list for the subscriber so that they know how to route messages (both requests and responses) to the user identity. Even if a hacker attempted a session hijack, for example, the proxies would ignore the routing provided in the message, relying instead on the route list they recorded for the subscriber during the registration.

The hacker then would have to hijack a registration and successfully register his or her own address using the subscriber identity. This is still possible, unless the network is using encryption and authentication keys, in which case the hacker would be unable to read the intercepted messages (thereby being prevented from capturing the user identity) and would be unable to pass through authentication (not having the proper authentication keys).

Strict routing does have its drawbacks, though. This is not a very efficient means of network routing because it forces traffic through specific paths regardless of network utilization at that specific time. This means the network will have load imbalance, and many facilities may sit idle during certain periods of the day.

This, of course, goes against the grain of IP networks, where routing is determined according to the network conditions. It does align more with the circuit-switched network procedures, so switching engineers will be very comfortable with this form of routing, but IT engineers will not be so thrilled about strict routing.

If used with all of the other approaches we have talked about so far, this will certainly make the network a very robust and well-protected network. Nothing is perfect, and there is no such thing as a totally secure network as we have seen demonstrated time and time again. But remember that the object to security is to make it difficult for the hacker to fool the network, while allowing your subscribers to enjoy an easy and feature-rich, yet secure network experience.

Security Solutions

There are many reasons why many networks do not implement any form of security. Many platforms are older systems running operating systems that do not support security patches and must be completely replaced by newer platforms. This of course is cost prohibitive.

Other operators are severely short-handed and lack both the resources and the expertise to implement a strong security policy. They also lack the capital to invest in security implementations. Human error and configuration mistakes add to the problem, especially when expertise is lacking.

There are many different types of solutions available for securing networks. The security industry is probably one of the fastest growing tech sectors today, with products ranging the full gambit. This is both good and bad. With so many different products available, there are also just as many different approaches to security.

One worthy change within security concepts today is the layering of security implementations. Layering means implementing a security solution at all layers of the network. There are many reasons why this approach makes a lot of sense. When looking at the various types of attacks, some attacks are best detected at the lower layers, while others can only be detected at the application layers.

For example, security implemented at the network level will detect network anomalies and changes in the traffic flow but cannot detect buffer overruns in the application servers. Systems that operate at the network level have no visibility to the applications themselves and therefore are unable to detect problems within the applications themselves.

A layered approach provides a lot more flexibility to the operator as well. It allows operators to scale their security implementation rather than invest everything at the transport layer. By implementing just what is needed at the various layers, operators can save on their overall investments significantly.

Here is why. I could launch a distributed denial of service (DDoS) attack on a network that would be very difficult to detect at the transport layer. This is because I am sending large volumes of *INVITEs* from many different origination points within the network using bots. I would need to collect data from all over the network to determine that there was an attack of this nature in progress, or rely on an alarm at the destination point when it went into congestion.

However, if I am monitoring the application layer, I will see an increase in traffic in real time. Furthermore, I will see that the traffic is coming from multiple origination points and looks to be a DDoS attack. I could invest more money to implement sophisticated analysis applications at the transport layer combined with deep packet inspection, or I could invest in detection software at the application level (and most likely not have to invest as much).

The concept of layering your implementations allows you to make the right amount of investment at each layer, without having to purchase very expensive solutions at one layer that does all. It also provides a much more robust platform, allowing you to provide various layers of security for different segments of the network and various applications and services.

There is no single-point security solution that will protect the entire network, so a layered approach is the best means of ensuring your network assets are protected. Also remember that a single security solution is purpose built, designed to protect against specific types of security breaches. All security plans should incorporate several different platforms and methods based on the type of network and the services being provided.

If the services consist mostly of content, then that content is what must be protected. Digital rights management (DRM) will be required to deter copyright violations, while security measures such as those discussed will be necessary to prevent unauthorized access to the network and the content.

The following sections discuss the various solutions available to protect a network from attack. These are not exclusive; there are numerous different solutions and approaches, but these should be considered as a bare minimum.

Intrusion Detection

Monitoring systems have been used for several years to monitor the health of the network. Now they bring additional value as intrusion detection systems (IDSs). Of course, traditional network monitoring only has visibility to the call control layers within your network. You will need a platform to provide additional visibility into the transport layers as well.

This is especially true of data-intensive networks where the primary service is data services rather than just voice. In a voice network, traditional monitoring systems can easily be implemented to detect specific attack scenarios such as ISUP flooding (in an SS7 network) or SIP *INVITE* flooding (in a SIP network).

An IDS can operate in real time (or near real time) or historical mode. A real-time system is needed for detecting attacks while they are in progress. However, these systems should also have some capacity of storage to allow for the investigation of network events at a later time. The amount of storage depends on the amount of traffic to be stored, and the duration of time you need to review.

For example, if you want to be able to pull traffic for a six-month period, you will obviously need storage capacity for six months of traffic (or possibly more). While this is not common today, there are more and more network operators who are implementing long-term storage of their traffic for investigating incidents (as well as traffic modeling, revenue assurance, lawful intercept, and many other applications).

The purpose of these intrusion detection systems is to detect and notify of certain network anomalies at the call control layer. They should have the ability to see all SIP methods, and they should have the ability to provide basic key performance indicators (KPIs). These KPIs later become important for identifying flooding scenarios and other attack methods.

The IDS identifies the source of data (the network, application, or host). It performs analysis on the traffic based on rules (policy) and could also have the ability to establish its own policy through neural technology. The IDS then sends notification of the event to a console or other reporting system.

For example, in a flooding case, the monitoring system should detect an increase in the number of SIP requests across the entire network, as well as specific network segments. This could indicate (if it is a sharp rise in the number of requests) that there is a DoS attack underway.

If the system also supports the configuration of thresholds, then the system can be set to alert the users anytime these thresholds are exceeded. This is important for identifying DoS attacks.

Another advantage to monitoring systems is the ability to measure the performance and set thresholds for the entire network, or critical network segments. Because the monitoring system has visibility to all network elements, and all network facilities (if implemented network wide), the system can therefore see traffic levels across the entire network rather than within just one entity.

If you are relying on alarms from the various network entities, you will most likely not see a distributed DoS attack until it's too late. This is because a DDoS attack comes from many different sources, originating from many UAs across the network. Without a view of the entire network, you will not be able to see all of the traffic.

The individual network elements will only be able to detect and provide data on what they see, which is only a fraction of the traffic. For example, security implemented within proxies is great for identifying situations at the proxy, but they will miss anything that happens within other proxies, or occurs at any other point in the network.

For this reason it is best to leave network element security implementations focused on transport security, and rely on monitoring systems for security at the call control layers. This is the best way to ensure you will see all situations that could take place at that layer.

Monitoring systems can be thought of as network-based IDSs. They use probes to capture the network traffic, based on rules (or filters) defining the type of traffic. Usually the filters support defining traffic specifics such as origination points, type of traffic, and so on.

The probes then report back to a central server where the final analysis is completed. This server will also provide some form of console for reporting incidents and generating alarms. The probes themselves are passive, although they can be active in some cases.

Because they are passive, they do not require a network address. This makes them transparent in the network, which adds another distinct advantage. Because they cannot be detected, they are invisible to hackers.

The network-based IDS is best deployed both at the network edges (monitoring the access points into the network) and at aggregation points within the network (where it will be able to detect man-in-the-middle and other internal attacks).

One disadvantage to passive probes is that they do not work in networks where encryption is used. Because they are passive, they are unable to actively exchange encryption keys and negotiate these keys with other network elements. This can be a major setback, since encryption is key to any network security strategy.

There are a number of integrated solutions for detecting network intrusions at the transport layer. Host-based IDSs are typically implemented within the proxies and the gateways in the network.

The IDS should be separate from the elements it is protecting. In other words, it is not desirable to integrate the IDS function within the various network elements, since an attack on the network element would also compromise the IDS. The exception to this rule is when you are using a combination of IDS implementations. When combined with network IDSs, host-based IDSs can provide an additional advantage.

A host-based IDS monitors the various ports as well as processes within each of the network elements. This provides another set of eyes at the network element itself that a probe would never be able to detect. This can be important especially when looking for probing events where hackers are scanning platforms looking for vulnerabilities.

Another advantage of a host-based IDS is that it works within an encrypted network. Since it resides within the various hosts, the host-based IDS does not need to know the encryption keys. It is dealing with internal processes within the host itself, and any traffic data it receives will already be decrypted by the host it resides in.

Host-based IDSs are also capable of detecting viruses and Trojan horses residing on the host. Only a host-based IDS would have this visibility, since it requires visibility to internal processes and ports on the host. This is also a major advantage and key reason for implementing host-based IDSs in conjunction with network-based IDSs.

With all of the advantages also come drawbacks. A host-based IDS still needs to report back to a central authority events going on within each of the hosts. A central console is needed to collect data from all the IDSs deployed and provide reports and alarms. Otherwise, operators would have to access each host individually and extract this information on a host-by-host basis.

Backhauling all of the data from multiple hosts can be a challenge and in some cases would require a lot of bandwidth. For this reason, not every host should be covered. Certainly the critical hosts within the network should be protected.

Host-based IDSs are also susceptible to attack, since they reside within the targets themselves. This can add to the challenge and, of course, if compromised will become worthless. A good example of this would be virus protection implemented on your computer being compromised so that it receives automatic updates from a hacker site rather than the software vendor's own site. This is a tactic used today by many hackers.

Since they reside in the hosts themselves, host-based IDSs obviously do not have visibility to the rest of the network. For this reason they should not be used as the sole security implementation within a network. They should always be used in conjunction network-based IDSs.

An application-based IDS is actually a subset of a host-based IDS. It resides on the host but monitors specific applications on the host rather than the whole host. The purpose of an application-based IDS is to provide visibility to a user's interaction with the application, detecting unauthorized use of an application.

The application-based IDS should always be used in conjunction with network and host-based IDSs, to provide a layered approach to security, as well as a total view of what is truly going on within the network and its resources. When implemented in this fashion, the overall IDS strategy can provide enough visibility for an operator to detect in real-time events in the network before they cause significant disruptions and outages.

An IDS performs analysis on collected data to determine if there is an attack underway, or abnormal events within the network. There are two types of methods that are used: signature analysis and anomaly analysis.

Signature analysis relies on rules that are defined within the IDS (also referred to as policy). These signatures are established from previous attacks, so they are signatures of known attacks used as profiles for detecting the same attacks again.

These are very accurate, since they are based on known attack signatures. They are good for networks where expertise may not be abundant or resources are limited, but they should not be the only analysis used. Signature analysis is based on known attacks and therefore is not well suited for finding new attacks that use a different signature.

Anomaly analysis looks for abnormal behavior in the network. This is the best method for finding new attack signatures, but it does require additional expertise, since the analysis is usually a manual process today.

Profiles are built based on a snapshot of captured traffic over a period of time. The longer the duration of time used to collect the traffic, the better the profile. One method may include setting thresholds in the network, and anytime these thresholds are exceeded, raising an alarm. Anything that deviates from the profile is then considered an anomaly.

Statistics can also be used for establishing profiles. Statistics such as key performance indicators (KPIs) and key quality indicators (KQIs) are commonly used today in many monitoring systems for analyzing traffic data and determining if there are security breaches.

Both methods should be used together for the most effective approach. A signature analysis implementation alone will not be effective in finding new attacks and will leave the network very vulnerable, since attack methods change regularly.

A new approach is to deploy an IDS as a "honeypot," where a host is set as a decoy in the network. The data within the host is of enough interest to draw hackers, but benign to the network operator (no harm can be done with the data). When the attacker accesses the host, information is gathered about the attacker and all aspects of the session.

This information is then used in other IDSs within the network to establish a profile and signature to be used to detect other network breaches. Sometimes the attacker is even transferred to another part of the host where he or she is isolated from the network and its resources, limiting the harm the attacker can render on the network.

In voice networks (especially wireless) there can be individual numbers within number ranges that are left unassigned (dark). These numbers are then closely monitored for activity. If there is an attempt to call one of these numbers, the call data is collected and analyzed, since it could be a hacker attempt to create a hit list within a network to be used as targets at a later time.

An IDS should be deployed using the layered approach discussed in this section, using all forms of IDS rather than just a single approach. If only one approach is feasible, however, the network-based approach is the best direction to take, since this will give you the best visibility to network activity.

Network probes should be deployed behind firewalls, as well as at network edges, on major backbones, and on critical subnets. This will give the best overall view of everything in the network. They can also be converted to include active capability at a later time (or at the time of implementation).

Intrusion Protection

An intrusion protection system (IPS) combines the analysis of an IDS with the added protection of a firewall. The IPS then must be configured with a network address, since it will be an active element within the network. It is able not only to alter received traffic but to generate traffic.

The IPS is capable of inspecting packets and altering packets, making them benign in the network. This allows rogue packets to continue in the network without sending failure responses back to the originators, alerting them that their attempt was not successful.

The IPS can be implemented as part of the IDS, or it can be implemented separately. In this type of implementation the IPS could receive instructions or data from the IDS or from a policy engine.

The IPS can also be integrated on a host with an application. Vendors are continually adding security features within their platforms to further enhance their applications. This adds another level of security, albeit at a slightly higher cost, since it must be implemented at all critical applications.

An IPS, like an IDS, must support the network protocols used within the network. This includes any vendor-proprietary protocols that are implemented on vendor platforms. It makes decisions based on policy, although in some cases intelligence can be added that allows the IPS to operate in a neural fashion, creating new profiles and policy based on historical traffic patterns. This requires storing traffic from many months for the most complete profile.

Many systems begin as passive IDSs and then later evolve into active IPSs. This is done by converting the passive probes to include active interfaces so that only those interfaces that are to be active would require a network address. This way, the passive capability can be maintained along with the active capability.

A

SIP-Related RFCs

There are of course many documents regarding SIP and SIP implementations. The IET accepts submissions from many different sources but ratifies only some of these submissions. It is the ratified RFCs that then become implemented by the operator community (and the vendor community building SIP products).

This appendix lists many of the main RFCs known to date that define SIP procedures and messaging. Many of these RFCs also identify extensions to be used in SIP as well, so it is important to understand which RFC defines which extensions. Many times there will be multiple RFCs for any one extension.

This is not meant to be an exhaustive list of IETF RFCs. As always, it is best to visit the IETF Web site to obtain the latest listing of RFCs. You can also visit the IANA Web site (www.iana.org) to view a listing of known extensions and messages, as well as their associated RFCs.

IETF SIP Requests for Comments (RFCs)

RFC 2327 SDP: Session Description Protocol

RFC 2486 The Network Access Identifier

RFC 2778 A Model for Presence and Instant Messaging

RFC 2779 Instant Messaging / Presence Protocol Requirements

RFC 2848 The PINT Service Protocol: Extensions to SIP and SDP for IP Access to Telephone Call Services

RFC 2976 The SIP INFO Method

RFC 3050 Common Gateway Interface for SIP

RFC 3087 Control of Service Context Using SIP Request-URI

RFC 3261 SIP: Session Initiation Protocol

RFC 3262 Reliability of Provisional Responses in the Session Initiation Protocol (SIP)

RFC 3263 Session Initiation Protocol (SIP): Locating SIP Servers

RFC 3265 Session Initiation Protocol (SIP)-Specific Event Notification

RFC 3311 The Session Initiation Protocol (SIP) UPDATE Method

RFC 3312 Integration of Resource Management and Session Initiation Protocol (SIP)

RFC 3313 Private Session Initiation Protocol (SIP) Extensions for Media Authorization

RFC 3315 The Session Initiation Protocol (SIP) Refer Method

RFC 3319 Dynamic Host Configuration Protocol (DHCPv6) Options for Session Initiation Protocol (SIP) Servers

RFC 3323 A Privacy Mechanism for the Session Initiation Protocol (SIP)

RFC 3324 Short Term Requirements for Network Asserted Identity

RFC 3325 Private Extensions to the Session Initiation Protocol (SIP) for Asserted Identity within Trusted Networks

RFC 3326 The Reason Header Field for the Session Initiation Protocol (SIP)

RFC 3327 Session Initiation Protocol (SIP) Extension Header Field for Registering Non-Adjacent Contacts

RFC 3329 Security Mechanism Agreement for the Session Initiation Protocol (SIP)

RFC 3361 Dynamic Host Configuration Protocol (DHCP-for-IPv4) Option for Session Initiation Protocol (SIP) Servers

RFC 3262 Reliability of Provisional Responses in the Session Initiation Protocol (SIP)

RFC 3263 Session Initiation Protocol (SIP): Locating SIP Servers

RFC 3264 An Offer/Answer Model with the Session Description Protocol (SDP)

RFC 3265 Session Initiation Protocol (SIP)-Specific Event Notification

RFC 3310 Hypertext Transfer Protocol (HTTP) Digest Authentication Using Authentication and Key Agreement (AKA)

RFC 3329 Security Mechanism Agreement for the Session Initiation Protocol (SIP)

RFC 3339 Date and Time on the Internet: Timestamps

RFC 3351 User Requirements for the Session Initiation Protocol (SIP) in Support of Deaf, Hard of Hearing and Speech-impaired Individuals

RFC 3372 Session Initiation Protocol for Telephones (SIP-T): Context and Architectures

RFC 3388 Grouping of Media Lines in the Session Description Protocol (SDP)

RFC 3398 Integrated Services Digital Network (ISDN) User Part (ISUP) to Session Initiation Protocol (SIP) Mapping

RFC 3420 Internet Media Type message/sipfrag

RFC 3428 Session Initiation Protocol (SIP) Extension for Instant Messaging

RFC 3455 Private Header (P-Header) Extensions to the Session Initiation Protocol (SIP) for the 3rd-Generation Partnership Project (3GPP)

RFC 3485 The Session Initiation Protocol (SIP) and Session Description Protocol (SDP) Static Dictionary for Signaling Compression (SigComp)

RFC 3486 Compressing the Session Initiation Protocol (SIP)

RFC 3487 Requirements for Resource Priority Mechanisms for the Session Initiation Protocol (SIP)

RFC 3515 The Session Initiation Protocol (SIP) Refer Method

RFC 3524 Mapping of Media Streams to Resource Reservation Flows

RFC 3539 Authentication, Authorization, and Accounting (AAA) Transport Profile

RFC 3578 Mapping of Integrated Services Digital Network (ISDN) User Part (ISUP) Overlap Signaling to the Session Initiation Protocol (SIP)

RFC 3581 An Extension to the Session Initiation Protocol (SIP) for Symmetric Response Routing

RFC 3588 Diameter Base Protocol

RFC 3603 Private Session Initiation Protocol (SIP) Proxy-to-Proxy Extensions for Supporting the PacketCable Distributed Call Signaling Architecture

RFC 3608 Session Initiation Protocol (SIP) Extension Header Field for Service Route Discovery During Registration

RFC 3665 Session Initiation Protocol (SIP) Basic Call Flow Examples

RFC 3666 Session Initiation Protocol (SIP) Public Switched Telephone Network (PSTN) Call Flows

RFC 3680 A Session Initiation Protocol (SIP) Event Package for Registrations

RFC 3702 Authentication, Authorization, and Accounting Requirements for the Session Initiation Protocol (SIP)

RFC 3725 Best Current Practices for Third Party Call Control (3pcc) in the Session Initiation Protocol (SIP)

RFC 3824 Using E.164 numbers with the Session Initiation Protocol (SIP)

RFC 3840 Indicating User Agent Capabilities in the Session Initiation Protocol (SIP)

RFC 3841 Caller Preferences for the Session Initiation Protocol (SIP)

RFC 3842 A Message Summary and Message Waiting Indication Event Package for the Session Initiation Protocol (SIP)

RFC 3891 The Session Initiation Protocol (SIP) "Replaces" Header

RFC 3892 The Session Initiation Protocol (SIP) Referred-By Mechanism

RFC 3903 Session Initiation Protocol (SIP) Extension for Event State Publication

RFC 3911 The Session Initiation Protocol (SIP) "Join" Header

RFC 3924 Cisco Architecture for Lawful Intercept in IP Networks

RFC 3959 The Early Session Disposition Type for the Session Initiation Protocol (SIP)

RFC 3960 Early Media and Ringing Tone Generation in the Session Initiation Protocol (SIP)

RFC 3969 The Internet Assigned Number Authority (IANA) Uniform Resource Identifier (URI) Parameter Registry for the Session Initiation Protocol (SIP)

RFC 4006 Diameter Credit-Control Application

RFC 4028 Session Timers in the Session Initiation Protocol (SIP)

RFC 4083 Input 3rd-Generation Partnership Project (3GPP) Release 5 Requirements on the Session Initiation Protocol (SIP)

RFC 4092 Usage of the Session Description Protocol (SDP) Alternative Network Address Types (ANAT) Semantics in the Session Initiation Protocol (SIP)

RFC 4122 A Universally Unique IDentifier (UUID) URN Namespace

RFC 4168 The Stream Control Transmission Protocol (SCTP) as a Transport for the Session Initiation Protocol (SIP)

RFC 4189 Requirements for End-to-Middle Security for the Session Initiation Protocol (SIP)

RFC 4244 An Extension to the Session Initiation Protocol (SIP) for Request History Information

RFC 4245 High-Level Requirements for Tightly Coupled SIP Conferencing

RFC 4411 Extending the Session Initiation Protocol (SIP) Reason Header for Preemption Events

RFC 4412 Communications Resource Priority for the Session Initiation Protocol (SIP)

RFC 4457 The Session Initiation Protocol (SIP) P-User-Database Private-Header (P-Header)

RFC 4474 Enhancements for Authenticated Identity Management in the Session Initiation Protocol (SIP)

RFC 4485 Guidelines for Authors of Extensions to the Session Initiation Protocol (SIP)

RFC 4488 Suppression of Session Initiation Protocol (SIP) REFER Method Implicit Subscription

RFC 4538 Request Authorization through Dialog Identification in the Session Initiation Protocol (SIP)

RFC 4662 A Session Initiation Protocol (SIP) Event Notification Extension for Resource Lists

RFC 4916 Connected Identity in the Session Initiation Protocol (SIP)

RFC 4964 The P-Answer-State Header Extension to the Session Initiation Protocol for the Open Mobile Alliance Push to Talk over Cellular

RFC 5002 The Session Initiation Protocol (SIP) P-Profile-Key Private Header (P-Header)

RFC 5009 Private Header (P-Header) Extension to the Session Initiation Protocol (SIP) for Authorization of Early Media

B

Methods and Parameters

This appendix identifies the parameters required for each method, depending on its usage by user agents. The use of these headers by a proxy differs and is covered in Appendix C. The tables in this appendix identify the various user agent (both UAC and UAS) scenarios where a method is used, and its contents.

There are many variables regarding the structure of each method, highly dependent on both implementation and the procedures being attempted. This appendix provides numerous examples but may not be exact in all cases. It is the intent to provide some examples of how each of the methods may look under these various circumstances as a guide.

The way to interpret this table is to look at the "Use" column to determine when the header field is used, and whether or not the field is optional or mandatory for that instance. The sending/receiving columns may appear confusing at first, but remember that a proxy may be the sending entity, in which case it may have mandatory headers that would be optional if a user agent were sending the same header.

For example, in the following table the *ACCEPT-CONTACT* header field is optional when being sent to communicate caller preferences. The *ALLOW-EVENTS* field, on the other hand, is used only when providing event notification and is mandatory if the entity is sending the field.

The values in this table are designated as follows:

- M = Mandatory
- O = Optional
- X = Prohibited
- I = Irrelevant
- C = Conditional

ACK Method

TABLE B.1   Supported Header Fields for ACK

Header Field	Sending to UA or Proxy	Receiving from UA or Proxy	When Used
Accept-Contact	O	N/A	Caller preferences
Allow-Events	O	M	Event notification
Authorization	M	M	Authentication between user agents
Call-ID	M	M	All instances of ACK
Content-Disposition	O	M	All instances of ACK
Content-Encoding	O	M	All instances of ACK
Content-Language	O	M	All instances of ACK
Content-Length	M	M	All instances of ACK
Content-Type	O	M	All instances of ACK
Cseq	M	M	All instances of ACK
Date	O	M	Insertion of date in requests and responses
From	M	M	All instances of ACK
Max-Forwards	M	N/A	All instances of ACK
MIME-Version	O	M	All instances of ACK
Privacy	O	O	Privacy mechanism
Proxy-Authorization	M	N/A	Authentication between user agent and proxy
Proxy-Require	O	N/A	All instances of ACK
Reason	O	O	Reason field
Reject-Contact	O	N/A	Caller preferences
Request-Disposition	O	N/A	Caller preferences
Require	O	M	All instances of ACK
Route	M	N/A	All instances of ACK
Timestamp	O	M	Timestamping of requests
To	M	M	All instances of ACK
User-Agent	O	M	All instances of ACK
Via	M	M	All instances of ACK

BYE Method

TABLE B.2 Supported Header Fields for BYE

Header	Sending to UA or Proxy	Receiving from UA or Proxy	When Used
Accept	O	M	All instances
Accept-Encoding	O	M	All instances
Accept-Language	O	M	All instances
Allow	O	M	All instances
Allow-Events	O	M	Event notification
Authorization	M	M	Authentication between UA and UA
Call-ID	M	M	All instances
Content-Disposition	O	M	All instances
Content-Encoding	O	M	All instances
Content-Language	O	M	All instances
Content-Length	M	M	All instances
Content-Type	M	M	All instances
Cseq	M	M	All instances
Date	O	M	Insertion of date in requests and responses
From	M	M	All instances
Max-Forwards	M	N/A	All instances
MIME-Version	O	M	All instances
P-Access-Network-Info	O	O	Charging network identification
P-Asserted-Identity	N/A	O	Asserted identity within trusted networks
P-Charging-Function-Addresses	O	O	Charging function address
P-Charging-Vector	O	O	Charging vector
P-Preferred-Identity	O	N/A	Preferred identity
Privacy	O	O	Privacy mechanism
Proxy-Authorization	M	N/A	Authentication between UA and proxy
Proxy-Require	O	N/A	All instances
Reason	O	O	Reason codes

(Continued)

TABLE B.2   Supported Header Fields for BYE (*continued*)

Header	Sending to UA or Proxy	Receiving from UA or Proxy	When Used
Record-Route	N/A	N/A	All instances
Reject-Contact	O	N/A	Caller preferences
Request-Disposition	O	N/A	Caller preferences
Require	O	M	All instances
Route	M	N/A	All instances
Security-Client	O	N/A	Security mechanisms in non-IMS networks
Security-Verify	M	N/A	Security mechanism
Supported	O	M	All instances
Timestamp	O	M	Timestamping of requests
To	M	M	All instances
User-Agent	O	O	All instances
Via	M	M	All instances

TABLE B.3   BYE Supported Header Fields When Used with 100 TRYING

Header	Sending to UA or Proxy	Receiving from UA or Proxy
Call-ID	N/A	M
Content-Length	N/A	M
Cseq	N/A	M
Date	N/A	M
From	N/A	M
To	N/A	M
Via	N/A	M

TABLE B.4   BYE Supported Header Fields When Used with 2*xx* Response

Header	Sending to UA or Proxy	Receiving from UA or Proxy	Use
Allow	O	M	
Authentication-Info	O	M	Authentication between UA and UA
Supported	M	M	

TABLE B.5   **BYE Supported Header Fields When Used with Response 300, 301, 302, 305, 380, or 485**

Header	Sending to UA or Proxy	Receiving from UA or Proxy
Allow	O	M
Contact	O	M
Error-Info	O	O
Supported	M	M

TABLE B.6   **BYE Supported Header Fields When Used with Response 401 Unauthorized**

Header	Sending to UA or Proxy	Receiving from UA or Proxy	Use
	RFC status	RFC status	
Allow	O	M	
Error-Info	O	O	
Proxy-Authenticate	M	M	Authentication between UA and UA
Supported	M	M	
WWW-Authenticate	M	M	

TABLE B.7   **BYE Supported Header Fields When Used with Response 404, 413, 480, 486, 500, 503, 600, or 603**

Header	Sending to UA or Proxy	Receiving from UA or Proxy
Allow	O	M
Error-Info	O	O
Retry-After	O	O
Supported	M	M

TABLE B.8   **BYE Supported Header Fields When Used with Response 405**

Header	Sending to UA or Proxy	Receiving from UA or Proxy
Allow	M	M
Error-Info	O	O
Supported	M	M

TABLE B.9   BYE Supported Header Fields When Used with Response 407

Header	Sending to UA or Proxy	Receiving from UA or Proxy	Use
Allow	O	M	
Error-Info	O	O	
Proxy-Authenticate	M	M	Authentication between UA and UA
Supported	M	M	
WWW-Authenticate	O	O	

TABLE B.10   BYE Supported Header Fields When Used with Response 415

Header	Sending to UA or Proxy	Receiving from UA or Proxy
Accept	O	M
Accept-Encoding	O	M
Accept-Language	O	M
Allow	O	M
Error-Info	O	O
Supported	M	M

TABLE B.11   BYE Supported Header Fields When Used with Response 420

Header	Sending to UA or Proxy	Receiving from UA or Proxy
Allow	O	M
Error-Info	O	O
Supported	M	M
Unsupported	M	M

TABLE B.12   BYE Supported Header Fields When Used with Response 421 or 494

Header	Sending to UA or Proxy	Receiving from UA or Proxy	Use
Allow	O	M	
Error-Info	O	O	
Security-Server	O	M	Security mechanism
Supported	M	M	

TABLE B.13   BYE Supported Header Fields When Used with Response 484

Header	Sending to UA or Proxy	Receiving from UA or Proxy
Allow	O	M
Error-Info	O	O
Supported	M	M

TABLE B.14   BYE Supported Header Fields When Used with All Other Responses

Header	Sending to UA or Proxy	Receiving from UA or Proxy
Call-ID	M	M
Content-Disposition	O	M
Content-Encoding	O	M
Content-Language	O	M
Content-Length	M	M
Content-Type	M	M
Cseq	M	M
Date	M	M
From	M	M
MIME-Version	O	M
P-Access-Network-Info	O	O
P-Asserted-Identity	N/A	O
P-Charging-Function-Addresses	O	O
P-Charging-Vector	O	O
P-Preferred-Identity	O	N/A
Privacy	O	O
Require	M	M
Server	O	O
Timestamp	M	M
To	M	M
User-Agent	O	O
Via	M	M
Warning	O	O

CANCEL Method

TABLE B.15   CANCEL Supported Header Fields

Header	Sending to UA or Proxy	Receiving from UA or Proxy	Use
Accept-Contact	O	N/A	Caller preferences
Allow-Events	O	M	Event notification
Authorization	M	M	Authentication between UA and UA
Call-ID	M	M	All instances
Content-Length	M	M	All instances
Cseq	M	M	All instances
Date	O	M	Insertion of date in requests and responses
From	M	M	All instances
Max-Forwards	M	N/A	All instances
Privacy	O	O	Privacy mechanism
Reason	O	O	Reason codes
Record-Route	N/A	N/A	All instances
Reject-Contact	O	N/A	Caller preferences
Request-Disposition	O	N/A	Caller preferences
Route	M	N/A	All instances
Supported	O	M	All instances
Timestamp	O	M	Timestamping of requests
To	M	M	All instances
User-Agent	O	O	All instances
Via	M	M	All instances

TABLE B.16   CANCEL Supported Header Fields When Used with Response 200

Header	Sending to UA or Proxy	Receiving from UA or Proxy
Record-Route	N/A	N/A
Supported	M	M

TABLE B.17   CANCEL Supported Header Fields When Used with Response 401

Header	Sending to UA or Proxy	Receiving from UA or Proxy
Error-Info	O	O
Supported	M	M

TABLE B.18   CANCEL Supported Header Fields When Used with Response 404, 413, 480, 500, 503, 600, or 603

Header	Sending to UA or Proxy	Receiving from UA or Proxy
Error-Info	O	O
Retry-After	O	O
Supported	M	M

TABLE B.19   CANCEL Supported Header Fields When Used with Response 484

Header	Sending to UA or Proxy	Receiving from UA or Proxy
Error-Info	O	O
Supported	M	M

TABLE B.20   CANCEL Supported Header Fields When Used with All Other Responses

Header	Sending to UA or Proxy	Receiving from UA or Proxy	Use
Call-ID	M	M	
Content-Length	M	M	
Cseq	M	M	
Date	O	M	Insertion of date in requests and responses
From	M	M	
Privacy	O	O	Privacy mechanism
Timestamp	M	M	Timestamping of requests
To	M	M	
User-Agent	O	O	
Via	M	M	
Warning	O	O	

INVITE Method

TABLE B.21   INVITE Supported Header Fields

Header	Sending to UA or Proxy	Receiving from UA or Proxy	Use
Accept	O	M	
Accept-Contact	O	N/A	Caller preferences
Accept-Encoding	O	M	
Accept-Language	O	M	
Alert-Info	O	M	Downloading of alert information
Allow	O	M	
Allow-Events	M	M	Event notification
Authorization	M	M	Authentication between UA and UA
Call-ID	M	M	
Call-Info	O	O	
Contact	M	M	
Content-Disposition	O	M	
Content-Encoding	O	M	
Content-Language	O	M	
Content-Length	M	M	
Content-Type	M	M	
Cseq	M	M	
Date	O	M	Insertion of date in requests and responses
Expires	O	O	
From	M	M	
In-Reply-To	O	O	
Max-Forwards	M	N/A	
MIME-Version	O	M	
Organization	O	O	
P-Access-Network-Info	O	O	Charging network information
P-Asserted-Identity	N/A	O	Asserted identity
P-Called-Party-ID	O	O	Called party identifier
P-Charging-Function-Addresses	O	O	Charging function address
P-Charging-Vector	O	O	Charging vector
P-Media-Authorization	N/A	M	Media authorization
P-Preferred-Identity	O	N/A	Preferred identity

TABLE B.21   INVITE Supported Header Fields (*continued*)

Header	Sending to UA or Proxy	Receiving from UA or Proxy	Use
P-Visited-Network-ID	O	O	Visited network identifier
Priority	O	O	
Privacy	O	O	Privacy mechanism
Proxy-Authorization	M	N/A	Authentication between proxy and UA
Proxy-Require	O	N/A	
Reason	O	O	Reason codes
Record-Route	N/A	M	
Reply-To	O	O	
Require	O	M	
Route	M	N/A	
Security-Client	O	N/A	Security mechanism
Security-Verify	M	N/A	Security mechanism
Subject	O	O	
Supported	O	M	Reason codes
Timestamp	O	M	Timestamping of requests
To	M	M	
User-Agent	O	O	
Via	M	M	

TABLE B.22   INVITE Supported Header Fields When Used with Response 100

Header	Sending to UA or Proxy	Receiving from UA or Proxy
Call-ID	N/A	M
Content-Length	N/A	M
Cseq	N/A	M
Date	N/A	M
From	N/A	M
To	N/A	M
Via	N/A	M

TABLE B.23 INVITE Supported Header Fields When Used with 1xx Response

Header	Sending to UA or Proxy	Receiving from UA or Proxy	Use
Allow	O	M	
Contact	O	M	
P-Media-Authorization	N/A	M	Media authorization
Supported	O	M	

TABLE B.24 INVITE Supported Header Fields When Used with 2xx Response

Header	Sending to UA or Proxy	Receiving from UA or Proxy	Use
Accept	O	M	
Accept-Encoding	O	M	
Accept-Language	O	M	
Allow	O	M	
Authentication-Info	O	M	Authentication between UA and UA
Contact	M	M	
P-Media-Authorization	N/A	M	Media authorization
Record-Route	M	M	
Expires	M	M	
Supported	M	M	

TABLE B.25 INVITE Supported Header Fields When Used with 3xx or 485 Response

Header	Sending to UA or Proxy	Receiving from UA or Proxy
Allow	O	M
Contact	O	M
Error-Info	O	O
Supported	M	M

TABLE B.26 INVITE Supported Header Fields When Used with Response 401

Header	Sending to UA or Proxy	Receiving from UA or Proxy	Use
Allow	O	M	
Error-Info	O	O	
Proxy-Authenticate	M	M	Authentication between UA and UA
Supported	M	M	
WWW-Authenticate	M	M	

TABLE B.27 INVITE Supported Header Fields When Used with Response 404, 413, 480, 486, 600, or 603

Header	Sending to UA or Proxy	Receiving from UA or Proxy
Allow	O	M
Error-Info	O	O
Retry-After	O	O
Supported	M	M

TABLE B.28 INVITE Supported Header Fields When Used with Response 405

Header	Sending to UA or Proxy	Receiving from UA or Proxy
Allow	M	M
Error-Info	O	O
Supported	M	M

TABLE B.29 INVITE Supported Header Fields When Used with Response 407

Header	Sending to UA or Proxy	Receiving from UA or Proxy
Allow	O	M
Error-Info	O	O
Proxy-Authenticate	O	O
Supported	M	M
WWW-Authenticate	O	O

TABLE B.30 INVITE Supported Header Fields When Used with Response 415

Header	Sending to UA or Proxy	Receiving from UA or Proxy
Accept	O	M
Accept-Encoding	O	M
Accept-Language	O	M
Allow	O	M
Error-Info	O	O
Supported	M	M

TABLE B.31   INVITE Supported Header Fields When Used with Response 420

Header	Sending to UA or Proxy	Receiving from UA or Proxy
Allow	O	M
Error-Info	O	O
Supported	M	M
Unsupported	M	M

TABLE B.32   INVITE Supported Header Fields When Used with Response 421 or 494

Header	Sending to UA or Proxy	Receiving from UA or Proxy	Use
Allow	O	M	
Error-Info	O	O	
Security-Server	O	M	Security mechanism
Supported	M	M	

TABLE B.33   INVITE Supported Header Fields When Used with Response 484

Header	Sending to UA or Proxy	Receiving from UA or Proxy
Allow	O	M
Error-Info	O	O
Supported	M	M

TABLE B.34   INVITE Supported Header Fields When Used with Response 500

Header	Sending to UA or Proxy	Receiving from UA or Proxy
Allow	O	M
Error-Info	O	O
Retry-After	M	O
Supported	M	M

TABLE B.35   INVITE Supported Header Fields When Used with Response 503

Header	Sending to UA or Proxy	Receiving from UA or Proxy
Allow	O	M
Error-Info	O	O
Retry-After	O	O
Supported	M	M

TABLE B.36   INVITE Supported Header Fields When Used with All Other Responses

Header	Sending to UA or Proxy	Receiving from UA or Proxy	Use
Call-ID	M	M	
Call-Info	O	O	
Content-Disposition	O	M	
Content-Encoding	O	M	
Content-Language	O	M	
Content-Length	M	M	
Content-Type	M	M	
Cseq	M	M	
Date	O	M	Insertion of date in requests and responses
From	M	M	
MIME-Version	O	M	
Organization	O	O	
P-Access-Network-Info	O	O	Access network identifier
P-Asserted-Identity	N/A	O	Asserted identity
P-Charging-Function-Addresses	O	M	Charging function address
P-Charging-Vector	O	O	Charging vector
P-Preferred-Identity	O	N/A	Preferred identity
Privacy	O	O	Privacy mechanism
Require	M	M	
Server	O	O	
Timestamp	M	M	Timestamping of requests
To	M	M	
User-Agent	O	O	
Via	M	M	
Warning	O	O	

MESSAGE Method

TABLE B.37   MESSAGE Supported Header Fields

Header	Sending to UA or Proxy	Receiving from UA or Proxy	Use
Accept-Contact	O	N/A	Caller preferences
Allow	O	M	
Allow-Events	O	M	Event notification
Authorization	M	M	Authentication between UA and UA
Call-ID	M	M	
Call-Info	O	O	
Content-Disposition	O	M	
Content-Encoding	O	M	
Content-Language	O	M	
Content-Length	M	M	
Content-Type	M	M	
Cseq	M	M	
Date	O	M	Insertion of date on requests and responses
Expires	O	O	
From	M	M	
In-Reply-To	O	O	
Max-Forwards	M	N/A	
MIME-Version	O	M	
Organization	O	O	
P-Access-Network-Info	O	O	Access network identifier
P-Asserted-Identity	N/A	O	Asserted identity
P-Called-Party-ID	O	O	Called party identifier
P-Charging-Function-Addresses	O	O	Charging function address
P-Charging-Vector	O	O	Charging vector identifier
P-Preferred-Identity	O	N/A	Preferred identity identifier
P-Visited-Network-ID	O	O	Visited network identifier
Priority	O	O	
Privacy	O	O	Privacy mechanism
Proxy-Authorization	M	N/A	Authentication between UA and proxy
Proxy-Require	O	N/A	

TABLE B.37   MESSAGE Supported Header Fields (*continued*)

Header	Sending to UA or Proxy	Receiving from UA or Proxy	Use
Reason	O	O	Reason codes
Record-Route	N/A	N/A	
Reply-To	O	O	
Require	O	M	Extensions required
Route	M	N/A	
Subject	O	O	
Supported	O	M	Extensions supported
Timestamp	O	M	Timestamping of requests
To	M	M	
User-Agent	O	O	
Via	M	M	

TABLE B.38   MESSAGE Supported Header Fields When Used with 2*xx* Response

Header	Sending to UA or Proxy	Receiving from UA or Proxy	Use
Allow	O	M	
Authentication-Info	O	M	Authentication between UA and UA
Supported	O	M	

TABLE B.39   MESSAGE Supported Header Fields When Used with 3*xx* or 485 Response

Header	Sending to UA or Proxy	Receiving from UA or Proxy
Allow	O	M
Contact	O	M
Error-Info	O	O
Supported	M	M

TABLE B.40   MESSAGE Supported Header Fields When Used with Response 401

Header	Sending to UA or Proxy	Receiving from UA or Proxy	Use
Allow	O	M	
Error-Info	O	O	
Proxy-Authenticate	O	O	Authentication between UA and UA
Supported	M	M	
WWW-Authenticate	M	M	

TABLE B.41   MESSAGE Supported Header Fields When Used with Response 404, 413, 480, 486, 500, 503, 600, or 603

Header	Sending to UA or Proxy	Receiving from UA or Proxy
Allow	O	M
Error-Info	O	O
Retry-After	O	O
Supported	M	M

TABLE B.42   MESSAGE Supported Header Fields When Used with Response 405

Header	Sending to UA or Proxy	Receiving from UA or Proxy
Allow	M	M
Error-Info	O	O
Supported	M	M

TABLE B.43   MESSAGE Supported Header Fields When Used with Response 415

Header	Sending to UA or Proxy	Receiving from UA or Proxy
Accept	O	M
Accept-Encoding	O	M
Accept-Language	O	M
Allow	O	M
Error-Info	O	O
Supported	M	M

TABLE B.44   MESSAGE Supported Header Fields When Used with Response 420

Header	Sending to UA or Proxy	Receiving from UA or Proxy
Allow	O	M
Error-Info	O	O
Supported	M	M
Unsupported	M	M

TABLE B.45   MESSAGE Supported Header Fields When Used with Response 421

Header	Sending to UA or Proxy	Receiving from UA or Proxy
Allow	O	M
Error-Info	O	O
Supported	M	M

TABLE B.46 MESSAGE Supported Header Fields When Used with Response 484

Header	Sending to UA or Proxy	Receiving from UA or Proxy
Allow	O	M
Error-Info	O	O
Supported	M	M

TABLE B.47 MESSAGE Supported Header Fields for All Other Responses

Header	Sending to UA or Proxy	Receiving from UA or Proxy	Use
Call-ID	M	M	
Call-Info	O	O	
Content-Disposition	O	M	
Content-Encoding	O	M	
Content-Language	O	M	
Content-Length	M	M	
Content-Type	M	M	
Cseq	M	M	
Date	O	M	Insert date in requests and responses
From	M	M	
MIME-Version	O	M	
Organization	O	O	
P-Access-Network-Info	O	O	Access network identifier
P-Asserted-Identity	N/A	O	Asserted identity
P-Charging-Function-Addresses	O	O	Charging function address
P-Charging-Vector	O	O	Charging vector identifier
P-Preferred-Identity	O	N/A	Preferred identity
Privacy	O	O	Privacy mechanism
Require	O	M	
Server	O	O	
Timestamp	M	O	Timestamping of responses
To	M	M	
User-Agent	O	O	
Via	M	M	
Warning	O	O	

NOTIFY Method

TABLE B.48   NOTIFY Supported Header Fields

Header	Sending to UA or Proxy	Receiving from UA or Proxy	Use
Accept	O	M	
Accept-Contact	O	N/A	Caller preferences
Accept-Encoding	O	M	
Accept-Language	O	M	
Allow	O	M	
Allow-Events	O	M	Caller preferences
Authorization	M	M	Authentication between UA and UA
Call-ID	M	M	
Contact	M	M	
Content-Disposition	O	M	
Content-Encoding	O	M	
Content-Language	O	M	
Content-Length	M	M	
Content-Type	M	M	
Cseq	M	M	
Date	O	M	Insertion of date in requests and responses
Event	M	M	
From	M	M	
Max-Forwards	M	N/A	
MIME-Version	O	M	
P-Access-Network-Info	O	O	Access network identifier
P-Asserted-Identity	N/A	O	Asserted identity
P-Charging-Function-Addresses	O	O	Charging function address
P-Charging-Vector	O	O	Charging vectors
P-Preferred-Identity	O	N/A	Preferred identity
Privacy	O	O	Privacy mechanism
Proxy-Authorization	M	N/A	Authentication between UA and proxy
Proxy-Require	O	N/A	
Reason	O	O	Reason codes

TABLE B.48 **NOTIFY Supported Header Fields (*continued*)**

Header	Sending to UA or Proxy	Receiving from UA or Proxy	Use
Record-Route	N/A	M	Strict routing
Reject-Contact	O	N/A	Caller preferences
Request-Disposition	O	N/A	Caller preferences
Require	O	M	
Security-Client	O	N/A	Security mechanism
Security-Verify	M	N/A	Security mechanism
Route	M	N/A	
Supported	O	M	
Timestamp	O	M	Timestamping requests
To	M	M	
User-Agent	O	O	
Via	M	M	

TABLE B.49 **NOTIFY Supported Header Fields When Used with 2*xx* Response**

Header	Sending to UA or Proxy	Receiving from UA or Proxy	Use
Allow	O	M	
Authentication-Info	O	M	Authentication between UA and UA
Contact	M	M	
Record-Route	M	M	Strict routing
Supported	M	M	

TABLE B.50 **NOTIFY Supported Header Fields When Used with Response 3*xx* or 485**

Header	Sending to UA or Proxy	Receiving from UA or Proxy
Allow	O	M
Contact	M	M
Error-Info	O	O
Supported	M	M

TABLE B.51 NOTIFY Supported Header Fields When Used with Response 401

Header	Sending to UA or Proxy	Receiving from UA or Proxy	Use
Allow	O	M	
Error-Info	O	O	
Proxy-Authenticate	M	M	Authentication between UA and UA
Supported	M	M	M
WWW-Authenticate	M	M	M

TABLE B.52 NOTIFY Supported Header Fields When Used with Response 404, 413, 480, 486, 500, 503, 600, or 603

Header	Sending to UA or Proxy	Receiving from UA or Proxy
Allow	O	M
Error-Info	O	O
Retry-After	O	O
Supported	M	M

TABLE B.53 NOTIFY Supported Header Fields When Used with Response 405

Header	Sending to UA or Proxy	Receiving from UA or Proxy
Allow	M	M
Error-Info	O	O
Supported	M	M

TABLE B.54 NOTIFY Supported Header Fields When Used with Response 407

Header	Sending to UA or Proxy	Receiving from UA or Proxy	Use
Allow	O	M	
Error-Info	O	O	
Proxy-Authenticate	M	M	Authentication between UA and UA
Supported	M	M	
WWW-Authenticate	O	O	

TABLE B.55 NOTIFY Supported Header Fields When Used with Response 415

Header	Sending to UA or Proxy	Receiving from UA or Proxy
Accept	O	M
Accept-Encoding	O	M
Accept-Language	O	M
Allow	O	M
Error-Info	O	O
Supported	M	M

TABLE B.56 NOTIFY Supported Header Fields When Used with Response 420

Header	Sending to UA or Proxy	Receiving from UA or Proxy
Allow	O	M
Error-Info	O	O
Supported	M	M
Unsupported	M	M

TABLE B.57 NOTIFY Supported Header Fields When Used with Response 421 or 494

Header	Sending to UA or Proxy	Receiving from UA or Proxy	Use
Allow	O	M	
Error-Info	O	O	
Security-Server	O	M	Security mechanism
Supported	M	M	

TABLE B.58 NOTIFY Supported Header Fields When Used with Response 484

Header	Sending to UA or Proxy	Receiving from UA or Proxy
Allow	O	M
Error-Info	O	O
Supported	M	M

TABLE B.59 NOTIFY Supported Header Fields When Used with Response 489

Header	Sending to UA or Proxy	Receiving from UA or Proxy
Allow	O	M
Allow-Events	M	M
Error-Info	O	O

TABLE B.60   NOTIFY Supported Header Fields with All Other Responses

Header	Sending to UA or Proxy	Receiving from UA or Proxy	Use
Call-ID	M	M	
Content-Disposition	O	M	
Content-Encoding	O	M	
Content-Language	O	M	
Content-Length	M	M	
Content-Type	M	M	
Cseq	M	M	
Date	O	M	Insertion of date in requests and responses
From	M	M	
MIME-Version	O	M	
P-Access-Network-Info	O	O	Access network identifier
P-Asserted-Identity	N/A	O	Asserted identity
P-Charging-Function-Addresses	O	O	Charging function address
P-Charging-Vector	O	O	Charging vector information
P-Preferred-Identity	O	N/A	Preferred identity
Privacy	O	O	Privacy mechanism
Require	M	M	
Server	O	O	
Timestamp	M	M	Timestamping of requests
To	M	M	
User-Agent	O	O	
Via	M	M	
Warning	O	O	

OPTIONS Method

TABLE B.61   OPTIONS Supported Header Fields

Header	Sending to UA or Proxy	Receiving from UA or Proxy	Use
Accept	M	M	
Accept-Contact	O	N/A	Caller preferences
Accept-Encoding	M	M	
Accept-Language	M	M	
Allow	O	M	
Allow-Events	M	M	Event notification
Authorization	M	M	Authentication between UA and UA

TABLE B.61 OPTIONS Supported Header Fields (*continued*)

Header	Sending to UA or Proxy	Receiving from UA or Proxy	Use
Call-ID	M	M	
Call-Info	O	O	
Contact	O	O	
Content-Disposition	O	M	
Content-Encoding	O	M	
Content-Language	O	M	
Content-Length	M	M	
Content-Type	M	M	
Cseq	M	M	
Date	O	M	Insertion of date in requests and responses
From	M	M	
Max-Forwards	M	N/A	
MIME-Version	O	M	
Organization	O	O	
P-Access-Network-Info	O	O	Access network identifier
P-Asserted-Identity	N/A	O	Asserted identity
P-Called-Party-ID	O	O	Called party identifier
P-Charging-Function-Addresses	O	O	Charging function address
P-Charging-Vector	O	O	Charging vector identifier
P-Preferred-Identity	O	N/A	Preferred identity
P-Visited-Network-ID	O	O	Visited network identifier
Privacy	O	O	Privacy mechanism
Proxy-Authorization	M	N/A	Authentication between UA and proxy
Proxy-Require	O	N/A	
Reason	O	O	Reason codes
Record-Route	N/A	N/A	
Request-Disposition	O	N/A	Caller preferences
Require	O	M	
Route	M	N/A	
Security-Client	O	N/A	Security mechanism
Security-Verify	O	N/A	Security mechanism
Supported	O	M	
Timestamp	O	M	Timestamping of requests
To	M	M	
User-Agent	O	O	
Via	M	M	

TABLE B.62   OPTIONS Supported Header Fields When Used with Response 100

Header	Sending to UA or Proxy	Receiving from UA or Proxy
Call-ID	N/A	M
Content-Length	N/A	M
Cseq	N/A	M
Date	N/A	M
From	N/A	M
To	N/A	M
Via	N/A	M

TABLE B.63   OPTIONS Supported Header Fields When Used with Response 2xx

Header	Sending to UA or Proxy	Receiving from UA or Proxy	Use
Accept	M	M	
Allow	O	M	
Authentication-Info	O	M	Authentication between UA and UA
Contact	O	O	
Supported	M	M	

TABLE B.64   OPTIONS Supported Header Fields When Used with Response 3xx or 485

Header	Sending to UA or Proxy	Receiving from UA or Proxy
Allow	O	M
Contact	O	M
Error-Info	O	O
Supported	M	M

TABLE B.65   OPTIONS Supported Header Fields When Used with Response 401

Header	Sending to UA or Proxy	Receiving from UA or Proxy	Use
Allow	O	M	
Error-Info	O	O	
Proxy-Authenticate	M	M	Authentication between UA and UA
Supported	M	M	
WWW-Authenticate	O	O	

TABLE B.66 OPTIONS Supported Header Fields When Used with Response 404, 413, 480, 486, 500, 503, 600, or 603

Header	Sending to UA or Proxy	Receiving from UA or Proxy
Allow	O	M
Error-Info	O	O
Retry-After	O	O
Supported	M	M

TABLE B.67 OPTIONS Supported Header Fields When Used with Response 405

Header	Sending to UA or Proxy	Receiving from UA or Proxy
Allow	M	M
Error-Info	O	O
Supported	M	M

TABLE B.68 OPTIONS Supported Header Fields When Used with Response 407

Header	Sending to UA or Proxy	Receiving from UA or Proxy	Use
Allow	O	M	
Error-Info	O	O	
Proxy-Authenticate	M	M	Authentication between UA and UA
Supported	M	M	
WWW-Authenticate	O	O	

TABLE B.69 OPTIONS Supported Header Fields When Used with Response 415

Header	Sending to UA or Proxy	Receiving from UA or Proxy
Accept	O	M
Accept-Encoding	O	M
Accept-Language	O	M
Allow	O	M
Error-Info	O	O
Supported	M	M

TABLE B.70 OPTIONS Supported Header Fields When Used with Response 420

Header	Sending to UA or Proxy	Receiving from UA or Proxy
Allow	O	M
Error-Info	O	O
Supported	M	M
Unsupported	M	M

TABLE B.71   OPTIONS Supported Header Fields When Used with Response 421 or 494

Header	Sending to UA or Proxy	Receiving from UA or Proxy	Use
Allow	O	M	
Error-Info	O	O	
Security-Server	O	M	Security mechanism
Supported	M	M	

TABLE B.72   OPTIONS Supported Header Fields When Used with Response 484

Header	Sending to UA or Proxy	Receiving from UA or Proxy
Allow	O	M
Error-Info	O	O
Supported	M	M

TABLE B.73   OPTIONS Supported Header Fields When Used with Response 484

Header	Sending to UA or Proxy	Receiving from UA or Proxy
Allow	O	M
Error-Info	O	O
Supported	M	M

TABLE B.74   OPTIONS Supported Header Fields When Used with All Other Responses

Header	Sending to UA or Proxy	Receiving from UA or Proxy	Use
Call-ID	M	M	
Call-Info	O	O	
Content-Disposition	O	M	
Content-Encoding	O	M	
Content-Language	O	M	
Content-Length	M	M	
Content-Type	M	M	
Cseq	M	M	
Date	O	M	Insertion of dates in requests and responses
From	M	M	
MIME-Version	O	M	

TABLE B.74 OPTIONS Supported Header Fields When Used with All Other Responses (*continued*)

Header	Sending to UA or Proxy	Receiving from UA or Proxy	Use
Organization	O	O	
P-Access-Network-Info	O	O	Access network identifier
P-Asserted-Identity	N/A	O	Asserted identity
P-Charging-Function-Addresses	O	O	Charging function address
P-Charging-Vector	O	O	Charging vector identifier
P-Preferred-Identity	O	N/A	Preferred identity
Privacy	O	O	Privacy mechanism
Require	M	M	
Timestamp	M	M	Timestamping of requests
To	M	M	
User-Agent	O	O	
Via	M	M	
Warning	O	O	

REGISTER Method

TABLE B.75 REGISTER Supported Header Fields

Header	Sending to UA or Proxy	Receiving from UA or Proxy	Use
Accept	O	M	
Accept-Encoding	O	M	
Accept-Language	O	M	
Allow	O	M	
Allow-Events	M	M	Event notification
Authorization	M	M	Authentication between UA and registrar
Call-ID	M	M	
Call-Info	O	O	
Contact	O	M	
Content-Disposition	O	M	
Content-Encoding	O	M	
Content-Language	O	M	

(*Continued*)

TABLE B.75   REGISTER Supported Header Fields (*continued*)

Header	Sending to UA or Proxy	Receiving from UA or Proxy	Use
Content-Length	M	M	
Content-Type	M	M	
Cseq	M	M	
Date	O	M	Insertion of date in requests and responses
Expires	O	M	
From	M	M	
Max-Forwards	M	N/A	
MIME-Version	O	M	
Organization	O	O	
P-Access-Network-Info	O	O	Access network identifier
P-Charging-Function-Addresses	O	O	Charging function address
P-Charging-Vector	O	O	Charging vector identifier
P-Visited-Network-ID	O	O	Visited network identifier
Path	O	M	Registration of non-adjacent contacts
Privacy	O	O	Privacy mechanism
Proxy-Authorization	M	N/A	Authentication between UA and proxy
Proxy-Require	O	N/A	
Reason	O	O	Reason codes
Request-Disposition	O	N/A	Caller preferences
Require	O	M	
Route	O	N/A	
Security-Client	O	N/A	Security mechanism
Security-Verify	M	M	Security mechanism
Supported	O	M	
Timestamp	M	M	Timestamping of requests
To	M	M	
User-Agent	O	O	
Via	M	M	

TABLE B.76   REGISTER Supported Header Fields When Used with Response 100

Header	Sending to UA or Proxy	Receiving from UA or Proxy
Call-ID	N/A	M
Content-Length	N/A	M
Cseq	N/A	M
Date	N/A	M
From	N/A	M
To	N/A	M
Via	N/A	M

TABLE B.77   REGISTER Supported Header Fields When Used with Response 2xx

Header	Sending to UA or Proxy	Receiving from UA or Proxy	Use
Accept	O	O	
Accept-Encoding	O	M	
Accept-Language	O	M	
Allow	O	M	
Authentication-Info	O	M	Authentication between UA and registrar
Contact	O	M	
P-Associated-URI	M	O	Associated URI
Path	M	O	Registering non-adjacent contacts
Service-Route	M	M	Service route discovery during registration
Supported	M	M	

TABLE B.78   REGISTER Supported Header Fields When Used with Response 3xx or 485

Header	Sending to UA or Proxy	Receiving from UA or Proxy
Allow	O	M
Contact	O	M
Error-Info	O	O
Supported	M	M

TABLE B.79   REGISTER Supported Header Fields When Used with Response 401

Header	Sending to UA or Proxy	Receiving from UA or Proxy	Use
Allow	O	M	
Error-Info	O	O	
Proxy-Authenticate	M	M	Authentication between UA and UA
Security-Server	O	N/A	
Supported	M	M	
WWW-Authenticate	M	M	

TABLE B.80   REGISTER Supported Header Fields When Used with Response 404, 413, 480, 500, 503, 600, or 603

Header	Sending to UA or Proxy	Receiving from UA or Proxy
Allow	O	M
Error-Info	O	O
Retry-After	O	O
Supported	M	M

TABLE B.81   REGISTER Supported Header Fields When Used with Response 405

Header	Sending to UA or Proxy	Receiving from UA or Proxy
Allow	M	M
Error-Info	O	O
Supported	M	M

TABLE B.82   REGISTER Supported Header Fields When Used with Response 407

Header	Sending to UA or Proxy	Receiving from UA or Proxy	Use
Allow	O	M	
Error-Info	O	O	
Proxy-Authenticate	M	M	Authentication between UA and UA
Supported	M	M	
WWW-Authenticate	O	O	

TABLE B.83 REGISTER Supported Header Fields When Used with Response 415

Header	Sending to UA or Proxy	Receiving from UA or Proxy
Accept	O	M
Accept-Encoding	O	M
Accept-Language	O	M
Allow	O	M
Error-Info	O	O
Supported	M	M

TABLE B.84 REGISTER Supported Header Fields When Used with Response 420

Header	Sending to UA or Proxy	Receiving from UA or Proxy
Allow	O	M
Error-Info	O	O
Supported	M	M
Unsupported	M	M

TABLE B.85 REGISTER Supported Header Fields When Used with Response 421

Header	Sending to UA or Proxy	Receiving from UA or Proxy	Use
Allow	O	M	
Error-Info	O	O	
Security-Server	M	M	Security mechanism
Supported	M	M	

TABLE B.86 REGISTER Supported Header Fields When Used with Response 423

Header	Sending to UA or Proxy	Receiving from UA or Proxy
Allow	O	M
Error-Info	O	O
Min-Expires	M	M
Supported	M	M

TABLE B.87 REGISTER Supported Header Fields When Used with Response 484

Header	Sending to UA or Proxy	Receiving from UA or Proxy
Allow	O	M
Error-Info	O	O
Supported	M	M

TABLE B.88   REGISTER Supported Header Fields When Used with All Other Status Codes

Header	Sending to UA or Proxy	Receiving from UA or Proxy	Use
Call-ID	M	M	
Call-Info	O	O	
Content-Disposition	O	M	
Content-Encoding	O	M	
Content-Language	O	M	
Content-Length	M	M	
Content-Type	M	M	
Cseq	M	M	
Date	O	M	Insertion of date in requests and responses
From	M	M	
MIME-Version	O	M	
Organization	O	O	
P-Access-Network-Info	O	O	Access network identifier
P-Charging-Function-Addresses	O	O	Charging function address
P-Charging-Vector	O	O	Charging vector identifier
Privacy	O	O	Privacy mechanism
Require	M	M	
Server	O	O	
Timestamp	M	M	Timestamping of requests
To	M	M	
User-Agent	O	O	
Via	M	M	
Warning	O	O	

SUBSCRIBE Method

TABLE B.89   SUBSCRIBE Supported Header Fields

Header	Sending to UA or Proxy	Receiving from UA or Proxy	Use
Accept	O	M	
Accept-Contact	O	N/A	Caller preferences
Accept-Encoding	O	M	
Accept-Language	O	M	
Allow	O	M	
Allow-Events	O	M	Event notification
Authorization	M	M	Authentication between UA and UA

TABLE B.89   SUBSCRIBE Supported Header Fields (*continued*)

Header	Sending to UA or Proxy	Receiving from UA or Proxy	Use
Call-ID	M	M	
Contact	M	M	
Content-Disposition	O	M	
Content-Encoding	O	M	
Content-Language	O	M	
Content-Length	M	M	
Content-Type	M	M	
Cseq	M	M	
Date	O	M	Insertion of date into requests and responses
Event	M	M	
Expires	O	M	
From	M	M	
Max-Forwards	M	N/A	
MIME-Version	O	M	
Organization	O	O	
P-Access-Network-Info	O	O	Access network identifier
P-Asserted-Identity	N/A	O	Asserted identity
P-Called-Party-ID	O	O	Called party identifier
P-Charging-Function-Addresses	O	O	Charging function address
P-Charging-Vector	O	O	Charging vector identifier
P-Preferred-Identity	O	N/A	Preferred identity
P-Visited-Network-ID	O	O	Visited network identifier
Privacy	O	O	Privacy mechanism
Proxy-Authorization	M	N/A	Authentication between UA and proxy
Proxy-Require	O	N/A	
Reason	O	O	Reason codes
Record-Route	N/A	M	
Reject-Contact	O	N/A	Caller preferences
Request-Disposition	O	N/A	Caller preferences
Require	O	M	
Route	M	N/A	
Security-Client	O	N/A	Security mechanism
Security-Verify	M	N/A	Security mechanism
Supported	O	M	
Timestamp	O	M	Timestamping of requests
To	M	M	
User-Agent	O	O	
Via	M	M	

TABLE B.90   SUBSCRIBE Supported Header Fields When Used with 2xx Response

Header	Sending to UA or Proxy	Receiving from UA or Proxy	Use
Allow	O	M	
Authentication-Info	O	M	Authentication between UA and UA
Contact	M	M	
Expires	M	M	
Require	M	M	
Supported	M	M	

TABLE B.91   SUBSCRIBE Supported Header Fields When Used with Response 3xx or 485

Header	Sending to UA or Proxy	Receiving from UA or Proxy
Allow	O	M
Contact	M	M
Error-Info	O	O
Supported	M	M

TABLE B.92   SUBSCRIBE Supported Header Fields When Used with Response 401

Header	Sending to UA or Proxy	Receiving from UA or Proxy	Use
Allow	O	M	
Error-Info	O	O	
Proxy-Authenticate	M	M	Authentication between UA and UA
Supported	M	M	
WWW-Authenticate	M	M	

TABLE B.93   SUBSCRIBE Supported Header Fields When Used with Response 404, 413, 480, 486, 500, 600, or 603

Header	Sending to UA or Proxy	Receiving from UA or Proxy
Allow	O	M
Error-Info	O	O
Retry-After	O	O
Supported	M	M

TABLE B.94   SUBSCRIBE Supported Header Fields When Used with Response 405

Header	Sending to UA or Proxy	Receiving from UA or Proxy
Allow	M	M
Error-Info	O	O
Supported	M	M

TABLE B.95   SUBSCRIBE Supported Header Fields When Used with Response 407

Header	Sending to UA or Proxy	Receiving from UA or Proxy	Use
Allow	O	M	
Error-Info	O	O	
Proxy-Authenticate	M	M	Authentication between UA and UA
Supported	M	M	
WWW-Authenticate	O	O	

TABLE B.96   SUBSCRIBE Supported Header Fields When Used with Response 415

Header	Sending to UA or Proxy	Receiving from UA or Proxy
Accept	O	M
Accept-Encoding	O	M
Accept-Language	O	M
Allow	O	M
Server	O	O
Supported	M	M

TABLE B.97   SUBSCRIBE Supported Header Fields When Used with Response 420

Header	Sending to UA or Proxy	Receiving from UA or Proxy
Allow	O	M
Error-Info	O	O
Supported	M	M
Unsupported	M	M

TABLE B.98   SUBSCRIBE Supported Header Fields When Used with Response 421 or 494

Header	Sending to UA or Proxy	Receiving from UA or Proxy	Use
Allow	O	M	
Error-Info	O	O	
Security-Server	O	M	Security mechanism
Supported	M	M	

TABLE B.99   SUBSCRIBE Supported Header Fields When Used with Response 423

Header	Sending to UA or Proxy	Receiving from UA or Proxy
Allow	O	M
Error-Info	O	O
Min-Expires	M	M
Supported	M	M

TABLE B.100   SUBSCRIBE Supported Header Fields When Used with Response 484

Header	Sending to UA or Proxy	Receiving from UA or Proxy
Allow	O	M
Error-Info	O	O
Supported	M	M

TABLE B.101   SUBSCRIBE Supported Header Fields When Used with Response 489

Header	Sending to UA or Proxy	Receiving from UA or Proxy
Allow	O	M
Allow-Events	M	M
Error-Info	O	O

TABLE B.102   SUBSCRIBE Supported Header Fields When Used with Response 503

Header	Sending to UA or Proxy	Receiving from UA or Proxy
Allow	O	M
Error-Info	O	O
Retry-After	O	O
Supported	M	M

TABLE B.103   SUBSCRIBE Supported Header Fields with All Other Responses

Header	Sending to UA or Proxy	Receiving from UA or Proxy	Use
Call-ID	M	M	
Content-Disposition	O	M	
Content-Encoding	O	M	
Content-Language	O	M	
Content-Length	M	M	
Content-Type	M	M	

TABLE B.103 SUBSCRIBE Supported Header Fields with All Other Responses (*continued*)

Header	Sending to UA or Proxy	Receiving from UA or Proxy	Use
Cseq	M	M	
Date	O	M	Insertion of dates in requests and responses
From	M	M	
MIME-Version	O	M	
Organization	O	O	
P-Access-Network-Info	O	O	Access network identifier
P-Asserted-Identity	N/A	O	Asserted identity
P-Charging-Function-Addresses	O	O	Charging function address
P-Charging-Vector	O	O	Charging vector identifier
P-Preferred-Identity	O	N/A	Preferred identity
Privacy	O	O	Privacy mechanism
Require	M	M	
Server	O	O	
Timestamp	M	M	Timestamping requests
To	M	M	
User-Agent	O	O	
Via	M	M	
Warning	O	O	

UPDATE Method

TABLE B.104 UPDATE Supported Header Fields

Header	Sending to UA or Proxy	Receiving from UA or Proxy	Use
Accept	O	M	
Accept-Contact	O	N/A	Caller preferences
Accept-Encoding	O	M	
Accept-Language	O	M	
Allow	O	M	
Allow-Events	O	M	Event notification
Authorization	M	M	Authentication between UA and UA

(*Continued*)

TABLE B.104   UPDATE Supported Header Fields (*continued*)

Header	Sending to UA or Proxy	Receiving from UA or Proxy	Use
Call-ID	M	M	
Call-Info	O	O	
Contact	M	M	
Content-Disposition	O	M	
Content-Encoding	O	M	
Content-Language	O	M	
Content-Length	M	M	
Content-Type	M	M	
Cseq	M	M	
Date	O	M	Insertion of dates in requests and responses
From	M	M	
Max-Forwards	M	N/A	
MIME-Version	O	M	
Organization	O	O	
P-Access-Network-Info	O	O	Access network identifier
P-Charging-Function-Addresses	O	O	Charging function address
P-Charging-Vector	O	O	Charging vector identifier
Privacy	O	O	Privacy mechanism
Proxy-Authorization	M	N/A	Authentication between UA and proxy
Proxy-Require	O	N/A	
Reason	O	O	Reason codes
Record-Route	N/A	N/A	
Request-Disposition	O	N/A	Caller preferences
Require	O	M	
Route	M	N/A	
Security-Client	O	N/A	Security mechanism
Security-Verify	M	N/A	Security mechanism
Expires	M	M	SIP session timer
Supported	O	M	
Timestamp	O	M	Timestamping of requests
To	M	M	
User-Agent	O	O	
Via	M	M	

TABLE B.105   UPDATE Supported Header Fields When Used with 2xx Response

Header	Sending to UA or Proxy	Receiving from UA or Proxy	Use
Accept	O	M	
Accept-Encoding	O	M	
Accept-Language	O	M	
Allow	O	M	
Authentication-Info	O	M	Authentication between UA and UA
Contact	M	M	
Expires	M	M	Session timer
Supported	M	M	

TABLE B.106   UPDATE Supported Header Fields When Used with 3xx Response

Header	Sending to UA or Proxy	Receiving from UA or Proxy
Allow	O	M
Contact	O	O
Error-Info	O	O
Supported	M	M

TABLE B.107   UPDATE Supported Header Fields When Used with Response 401

Header	Sending to UA or Proxy	Receiving from UA or Proxy
Allow	O	M
Error-Info	O	O
Proxy-Authenticate	O	O
Supported	M	M
WWW-Authenticate	M	M

TABLE B.108   UPDATE Supported Header Fields When Used with Response 404, 413, 480, 486, 500, 503, 600, or 603

Header	Sending to UA or Proxy	Receiving from UA or Proxy
Allow	O	M
Error-Info	O	O
Retry-After	O	O
Supported	M	M

TABLE B.109   UPDATE Supported Header Fields When Used with Response 405

Header	Sending to UA or Proxy	Receiving from UA or Proxy
Allow	M	M
Error-Info	O	O
Supported	M	M

TABLE B.110   UPDATE Supported Header Fields When Used with Response 407

Header	Sending to UA or Proxy	Receiving from UA or Proxy	Use
Allow	O	M	
Error-Info	O	O	
Proxy-Authenticate	O	O	Authentication between UA and UA
Supported	M	M	
WWW-Authenticate	O	O	

TABLE B.111   UPDATE Supported Header Fields When Used with Response 415

Header	Sending to UA or Proxy	Receiving from UA or Proxy
Accept	O	M
Accept-Encoding	O	M
Accept-Language	O	M
Allow	O	M
Error-Info	O	O
Supported	M	M

TABLE B.112   UPDATE Supported Header Fields When Used with Response 420

Header	Sending to UA or Proxy	Receiving from UA or Proxy
Allow	O	M
Error-Info	O	O
Supported	M	M
Unsupported	M	M

TABLE B.113   UPDATE Supported Header Fields When Used with Response 421 or 494

Header	Sending to UA or Proxy	Receiving from UA or Proxy	Use
Allow	O	M	
Error-Info	O	O	
Security-Server	O	M	Security mechanism
Supported	M	M	

TABLE B.114 **UPDATE Supported Header Fields When Used with All Other Responses**

Header	Sending to UA or Proxy	Receiving from UA or Proxy	Use
Call-ID	M	M	
Call-Info	O	O	
Content-Disposition	O	M	
Content-Encoding	O	M	
Content-Language	O	M	
Content-Length	M	M	
Content-Type	M	M	
Cseq	M	M	
Date	O	M	Insertion of dates in requests and responses
From	M	M	
MIME-Version	O	M	
Organization	O	O	
P-Access-Network-Info	O	O	Access network identifier
P-Charging-Function-Addresses	O	O	Charging function address
P-Charging-Vector	O	O	Charging vector identifier
Privacy	O	O	Privacy mechanism
Require	M	M	
Server	O	O	
Timestamp	M	M	Timestamping of requests
To	M	M	
User-Agent	O	O	
Via	M	M	
Warning	O	O	

C

Methods and Parameters from a Proxy Perspective

This appendix identifies the various parameters and their uses from a proxy perspective. This chapter is similar to Appendix B, for the exception that the parameters and methods identified here are based on use by a proxy.

The tables provide some examples as guidelines. There are many variables, and therefore the tables are not intended to be absolute by any means. Always refer to the RFCs for the most up-to-date requirements and usage.

The values in this table are designated as follows:

- M = Mandatory
- O = Optional
- X = Prohibited
- I = Irrelevant
- C = Conditional

ACK Method

TABLE C.1   ACK Supported Header Fields

Header	Sending to UA	Receiving from UA	Use
Accept-Contact	M	I	Caller preferences
Allow-Events	M	M	Event notification
Authorization	M	I	
Call-ID	M	M	
Content-Disposition	M	I	
Content-Encoding	M	I	
Content-Language	M	I	
Content-Length	M	M	
Content-Type	M	I	
Cseq	M	M	
Date	M	M	Insertion of dates in requests and responses
From	M	M	
Max-Forwards	M	M	
MIME-Version	M	I	
Privacy	M	M	Privacy mechanism
Proxy-Authorization	M	M	Authentication between UA and proxy
Proxy-Require	M	M	
Reason	M	I	Reason codes
Request-Disposition	M	I	Caller preferences
Require	M	M	Reading, adding, or modifying contents of REQUIRE header
Route	M	M	
Timestamp	M	I	
To	M	M	
User-Agent	M	I	
Via	M	M	

BYE Method

TABLE C.2   BYE Supported Header Fields

Header	Sending to UA	Receiving from UA	Use
Accept	M	I	
Accept-Contact	M	I	Caller preferences
Accept-Encoding	M	I	
Accept-Language	M	I	
Allow	M	I	
Allow-Events	M	M	Event notification
Authorization	M	I	
Call-ID	M	M	
Content-Disposition	M	I	
Content-Encoding	M	I	
Content-Language	M	I	
Content-Length	M	M	
Content-Type	M	I	
Cseq	M	M	
Date	M	M	Insertion of dates in requests and responses
From	M	M	
Max-Forwards	M	M	
MIME-Version	M	I	
P-Access-Network-Info	X	M	Access network identifier
P-Asserted-Identity	M	M	Asserted identity
P-Charging-Function-Addresses	M	M	Charging function identifier
P-Charging-Vector	M	M	Charging vector identifier
P-Preferred-Identity	X	M	Preferred identity
Privacy	M	M	Privacy mechanism
Proxy-Authorization	M	M	Authentication between UA and proxy
Proxy-Require	M	M	
Reason	M	I	Reason codes
Record-Route	M	O	Identifying proxy address

(Continued)

TABLE C.2   BYE Supported Header Fields (*continued*)

Header	Sending to UA	Receiving from UA	Use
Request-Disposition	M	I	Caller preferences
Require	M	M	Reading, adding, or modifying contents of REQUIRE header
Route	M	M	
Security-Client	X	M	Security mechanism
Security-Verify	X	M	Security mechanism
Supported	M	M	Reading contents of SUPPORTED header
Timestamp	M	I	
To	M	M	
User-Agent	M	I	
Via	M	M	

TABLE C.3   BYE Supported Header Fields Used with Response 100

Header	Sending to UA	Receiving from UA	Use
Call-ID	M	M	
Content-Length	M	M	
Cseq	M	M	
Date	M	M	Insertion of date in requests and responses
From	M	M	
To	M	M	
Via	M	M	

TABLE C.4   BYE Supported Header Fields Used with 2*xx* Response

Header	Sending to UA	Receiving from UA	Use
Allow	M	I	
Authentication-Info	M	I	
Record-Route	M	O	Adding proxy address
Supported	M	I	

TABLE C.5   **BYE Supported Header Fields Used with Response 3xx or 485**

Header	Sending to UA	Receiving from UA	Use
Allow	M	I	
Contact	M	M	Deleting CONTACT headers
Error-Info	M	I	
Supported	M	I	

TABLE C.6   **BYE Supported Header Fields Used with Response 401**

Header	Sending to UA	Receiving from UA
Allow	M	I
Error-Info	M	I
Proxy-Authenticate	M	M
Supported	M	I
WWW-Authenticate	M	I

TABLE C.7   **BYE Supported Header Fields Used with Response 404, 413, 480, 486, 500, 503, 600, or 603**

Header	Sending to UA	Receiving from UA
Allow	M	I
Error-Info	M	I
Retry-After	M	I
Supported	M	I

TABLE C.8   **BYE Supported Header Fields Used with Response 405**

Header	Sending to UA	Receiving from UA
Allow	M	I
Error-Info	M	I
Supported	M	I

TABLE C.9   **BYE Supported Header Fields Used with Response 407**

Header	Sending to UA	Receiving from UA
Allow	M	I
Error-Info	M	I
Proxy-Authenticate	M	M
Supported	M	I
WWW-Authenticate	M	I

TABLE C.10   BYE Supported Header Fields Used with Response 415

Header	Sending to UA	Receiving from UA
Accept	M	I
Accept-Encoding	M	I
Accept-Language	M	I
Allow	M	I
Error-Info	M	I
Supported	M	I

TABLE C.11   BYE Supported Header Fields Used with Response 420

Header	Sending to UA	Receiving from UA	Use
Allow	M	I	
Error-Info	M	I	
Supported	M	I	
Unsupported	M	M	When responding to other than REGISTER

TABLE C.12   BYE Supported Header Fields Used with Response 421 or 494

Header	Sending to UA	Receiving from UA	Use
Allow	O	M	
Error-Info	O	O	
Security-Server	M	N/A	Security mechanism
Supported	M	M	

TABLE C.13   BYE Supported Header Fields Used with Response 484

Header	Sending to UA	Receiving from UA
Allow	M	I
Error-Info	M	I
Supported	M	I

TABLE C.14   BYE Supported Header Fields Used with All Other Responses

Header	Sending to UA	Receiving from UA	Use
Call-ID	M	M	
Content-Disposition	M	I	
Content-Encoding	M	I	
Content-Language	M	I	
Content-Length	M	M	

TABLE C.14 BYE Supported Header Fields Used with All Other Responses (*continued*)

Header	Sending to UA	Receiving from UA	Use
Content-Type	M	I	
Cseq	M	M	
Date	M	M	Insertion of date in request and responses
From	M	M	
MIME-Version	M	I	
P-Access-Network-Info	X	M	Access network identifier
P-Asserted-Identity	M	M	Asserted identity
P-Charging-Function-Addresses	M	M	Charging function address
P-Charging-Vector	M	M	Charging vector identifier
P-Preferred-Identity	X	M	Preferred identity
Privacy	M	M	Privacy mechanism
Require	M	M	Registration process
Server	M	I	
Timestamp	M	I	
To	M	M	
User-Agent	M	I	
Via	M	M	
Warning	M	I	

CANCEL Method

TABLE C.15 CANCEL Supported Header Fields

Header	Sending to UA	Receiving from UA	Use
Accept-Contact	M	I	Caller preferences
Allow-Events	M	M	Event notification
Authorization	M	I	
Call-ID	M	M	
Content-Length	M	M	
Cseq	M	M	
Date	M	M	Insertion of date in requests and responses
From	M	M	

(*Continued*)

TABLE C.15   CANCEL Supported Header Fields (*continued*)

Header	Sending to UA	Receiving from UA	Use
Max-Forwards	M	M	
Privacy	M	M	Privacy mechanism
Reason	M	I	Reason codes
Record-Route	M	O	Insertion of proxy address for routing
Reject-Contact	M	I	Caller preferences
Request-Disposition	M	I	Caller preferences
Route	M	M	
Supported	M	M	Identifying methods supported
Timestamp	M	I	
To	M	M	
User-Agent	M	I	
Via	M	M	

TABLE C.16   CANCEL Supported Header Fields Used with Response 200

Header	Sending to UA	Receiving from UA	Use
Record-Route	M	M	Insertion of proxy address for routing
Supported	M	I	

TABLE C.17   CANCEL Supported Header Fields Used with Response 401

Header	Sending to UA	Receiving from UA
Error-Info	M	I
Supported	M	I

TABLE C.18   CANCEL Supported Header Fields Used with Response 404, 413, 480, 500, 503, 600, or 603

Header	Sending to UA	Receiving from UA
Error-Info	M	I
Retry-After	M	I
Supported	M	I

TABLE C.19   CANCEL Supported Header Fields Used with Response 484

Header	Sending to UA	Receiving from UA
Error-Info	M	I
Supported	M	I

TABLE C.20   CANCEL Supported Header Fields Used with All Other Responses

Header	Sending to UA	Receiving from UA	Use
Call-ID	M	M	
Content-Length	M	M	
Cseq	M	M	
Date	M	M	Insertion of date in requests and responses
From	M	M	
Privacy	M	M	Privacy mechanism
Timestamp	M	I	
To	M	M	
User-Agent	O	O	
Via	M	M	
Warning	M	I	

INVITE Method

TABLE C.21   INVITE Supported Header Fields

Header	Sending to UA	Receiving from UA	Use
Accept	M	I	
Accept-Contact	M	M	Caller preferences
Accept-Encoding	M	I	
Accept-Language	M	I	
Alert-Info	N/A	M	Alerting information
Allow	M	I	
Allow-Events	M	M	Event notification
Authorization	M	I	
Call-ID	M	M	
Contact	M	I	
Content-Disposition	M	I	
Content-Encoding	M	I	
Content-Language	M	I	
Content-Length	M	M	
Content-Type	M	I	
Cseq	M	M	
Date	M	M	Insertion of date in requests and responses

(Continued)

TABLE C.21   INVITE Supported Header Fields (*continued*)

Header	Sending to UA	Receiving from UA	Use
Expires	M	I	
From	M	M	
In-Reply-To	M	I	
Max-Forwards	M	M	
MIME-Version	M	I	
Min-SE	O	O	
Organization	M	M	Organization identifier
P-Access-Network-Info	X	M	Access network identifier
P-Asserted-Identity	M	M	Asserted identity
P-Called-Party-ID	M	I	Called party identifier
P-Charging-Function-Addresses	M	M	Charging function addresses
P-Charging-Vector	M	M	Charging vector identifier
P-Preferred-Identity	X	M	Preferred identity
P-Visited-Network-ID	M	M	Visited network identifier
Priority	M	I	
Privacy	M	M	Privacy mechanism
Proxy-Authorization	M	M	Authentication
Proxy-Require	M	M	
Reason	M	I	Reason codes
Record-Route	M	M	Inserting proxy address for routing
Reply-To	M	I	
Request-Disposition	M	M	Caller preferences
Require	M	M	Registration
Route	M	M	
Security-Client	X	M	Security mechanism
Security-Verify	X	M	Security mechanism
Expires	M	M	Session expiration timer
Subject	M	I	
Supported	M	M	Authorization
Timestamp	M	I	
To	M	M	
User-Agent	M	I	
Via	M	M	

TABLE C.22   INVITE Supported Header Fields Used with Response 100

Header	Sending to UA	Receiving from UA	Use
Call-ID	M	M	
Content-Length	M	M	
Cseq	M	M	
Date	M	I	Insertion of date in requests and responses
From	M	M	
To	M	M	
Via	M	M	

TABLE C.23   INVITE Supported Header Fields Used with 1xx Response

Header	Sending to UA	Receiving from UA
Allow	M	I
Contact	M	I
Rseq	M	I
Supported	M	I

TABLE C.24   INVITE Supported Header Fields Used with 2xx Response

Header	Sending to UA	Receiving from UA	Use
Accept	M	I	
Accept-Encoding	M	I	
Accept-Language	M	I	
Allow	M	I	
Authentication-Info	M	I	
Contact	M	I	
Record-Route	M	M	Insertion of proxy address for routing
Expires	M	M	Session expiration timer
Supported	M	I	

TABLE C.25   INVITE Supported Header Fields Used with Response 3xx or 485

Header	Sending to UA	Receiving from UA	Use
Allow	M	I	
Contact	M	M	Contact addresses
Error-Info	M	I	
Supported	M	I	

TABLE C.26   INVITE Supported Header Fields Used with Response 401

Header	Sending to UA	Receiving from UA
Allow	M	I
Error-Info	M	I
Proxy-Authenticate	M	M
Supported	M	I
WWW-Authenticate	O	O

TABLE C.27   INVITE Supported Header Fields Used with Response 404, 413, 480, 486, 600, or 603

Header	Sending to UA	Receiving from UA
Allow	M	I
Error-Info	M	I
Retry-After	M	I
Supported	M	I
Via	M	M

TABLE C.28   INVITE Supported Header Fields Used with Response 405

Header	Sending to UA	Receiving from UA
Allow	M	M/O
Error-Info	M	I
Supported	M	I

TABLE C.29   INVITE Supported Header Fields Used with Response 407

Header	Sending to UA	Receiving from UA
Allow	M	I
Error-Info	M	I
Proxy-Authenticate	M	M
Supported	M	I
WWW-Authenticate	M	I

TABLE C.30   INVITE Supported Header Fields Used with Response 415

Header	Sending to UA	Receiving from UA
Accept	M	I
Accept-Encoding	M	I
Accept-Language	M	I
Allow	M	I
Error-Info	M	I
Supported	M	I

TABLE C.31   INVITE Supported Header Fields Used with Response 420

Header	Sending to UA	Receiving from UA	Use
Allow	M	I	
Error-Info	M	I	
Supported	M	I	
Unsupported	M	M	Reading contents of UNSUPPORTED header prior to proxying

TABLE C.32   INVITE Supported Header Fields Used with Response 421 or 494

Header	Sending to UA	Receiving from UA	Use
Allow	O	M	
Error-Info	O	O	
Security-Server	M	N/A	Security mechanism
Supported	M	M	

TABLE C.33   INVITE Supported Header Fields Used with Response 484

Header	Sending to UA	Receiving from UA
Allow	M	I
Error-Info	M	I
Supported	M	I

TABLE C.34   INVITE Supported Header Fields Used with Response 500

Header	Sending to UA	Receiving from UA
Allow	M	I
Error-Info	M	I
Retry-After	M	I
Supported	M	I

TABLE C.35   INVITE Supported Header Fields Used with Response 503

Header	Sending to UA	Receiving from UA
Allow	M	I
Error-Info	M	I
Retry-After	M	I
Supported	M	I

TABLE C.36　INVITE Supported Header Fields Used with All Other Responses

Header	Sending to UA	Receiving from UA	Use
Call-ID	M	M	
Call-Info	M	M	Call-info for CSCF
Content-Disposition	M	I	
Content-Encoding	M	I	
Content-Language	M	I	
Content-Length	M	M	
Content-Type	M	I	
Cseq	M	M	
Date	M	M	Insertion of date in requests and responses
From	M	M	
MIME-Version	M	I	
Organization	M	M	Organization identifier
P-Access-Network-Info	M	I	Access network identifier
P-Asserted-Identity	M	M	Asserted identity
P-Charging-Function-Addresses	M	M	Charging function identifier
P-Charging-Vector	M	M	Charging vector identifier
P-Preferred-Identity	X	M	Preferred identity
Privacy	M	M	Privacy mechanism
Require	M	M	Required methods
Server	M	I	
Timestamp	M	I	
To	M	M	
User-Agent	M	I	
Via	M	M	
Warning	M	I	

MESSAGE Method

TABLE C.37　MESSAGE Supported Header Fields

Header	Sending to UA	Receiving from UA	Use
Allow	M	I	
Allow-Events	M	M	Event notification
Authorization	M	I	
Call-ID	M	M	
Call-Info	M	M	Call information
Content-Disposition	M	I	

TABLE C.37   MESSAGE Supported Header Fields (*continued*)

Header	Sending to UA	Receiving from UA	Use
Content-Encoding	M	I	
Content-Language	M	I	
Content-Length	M	M	
Content-Type	M	I	
Cseq	M	M	
Date	M	M	Insertion of dates in requests and responses
Expires	M	I	
From	M	M	
In-Reply-To	M	I	
Max-Forwards	M	M	
MIME-Version	M	I	
Organization	M	M	Organization identifier
P-Access-Network-Info	X	M	Access network identifier
P-Asserted-Identity	M	M	Asserted identity
P-Called-Party-ID	M	I	Called part identifier
P-Charging-Function-Addresses	M	M	Charging function addresses
P-Charging-Vector	M	M	Charging vector identifier
P-Preferred-Identity	X	M	Preferred identity
P-Visited-Network-ID	M	M	Visited network identifier
Priority	M	I	
Privacy	M	M	Privacy mechanism
Proxy-Authorization	M	M	Authentication between UA and proxy
Proxy-Require	M	M	
Reason	M	I	Reason codes
Record-Route	M	O	Insertion of proxy address for routing
Reply-To	M	I	
Request-Disposition	M	M	Caller preferences
Require	M	M	Required methods for the session
Route	M	M	
Security-Client	X	M	Security mechanism
Security-Verify	X	M	Security mechanism
Subject	M	I	
Supported	M	M	Supported methods by the proxy
Timestamp	M	I	
To	M	M	
User-Agent	M	I	
Via	M	M	

TABLE C.38   MESSAGE Supported Header Fields Used with 2xx Response

Header	Sending to UA	Receiving from UA	Use
Allow	M	I	
Authentication-Info	M	I	
Record-Route	M	O	Insertion of proxy address for routing
Supported	M	I	

TABLE C.39   MESSAGE Supported Header Fields Used with 3xx or 485 Response

Header	Sending to UA	Receiving from UA	Use
Allow	M	I	
Contact	M	M	Contact addresses
Error-Info	M	I	
Supported	M	I	

TABLE C.40   MESSAGE Supported Header Fields Used with Response 401

Header	Sending to UA	Receiving from UA
Allow	M	I
Error-Info	M	I
Proxy-Authenticate	M	M
Supported	M	I
WWW-Authenticate	M	I

TABLE C.41   MESSAGE Supported Header Fields Used with Response 404, 413, 480, 486, 500, 503, 600, or 603

Header	Sending to UA	Receiving from UA
Allow	M	I
Error-Info	M	I
Retry-After	M	I
Supported	M	I

TABLE C.42   MESSAGE Supported Header Fields Used with Response 405

Header	Sending to UA	Receiving from UA
Allow	M	I
Error-Info	M	I
Supported	M	I

TABLE C.43   MESSAGE Supported Header Fields Used with Response 407

Header	Sending to UA	Receiving from UA
Allow	M	I
Error-Info	M	I
Proxy-Authenticate	M	M
Supported	M	I
WWW-Authenticate	M	I

TABLE C.44   MESSAGE Supported Header Fields Used with Response 415

Header	Sending to UA	Receiving from UA
Accept	M	I
Accept-Encoding	M	I
Accept-Language	M	I
Allow	M	I
Error-Info	M	I
Supported	M	I

TABLE C.45   MESSAGE Supported Header Fields Used with Response 420

Header	Sending to UA	Receiving from UA	Use
Allow	M	I	
Error-Info	M	I	
Supported	M	I	
Unsupported	M	M	Reading contents prior to proxying a 420 response

TABLE C.46   MESSAGE Supported Header Fields Used with Response 421 or 494

Header	Sending to UA	Receiving from UA	Use
Allow	O	M	
Error-Info	O	O	
Security-Server	M	N/A	Security mechanism
Supported	M	M	

TABLE C.47   MESSAGE Supported Header Fields Used with Response 484

Header	Sending to UA	Receiving from UA
Allow	M	I
Error-Info	M	I
Supported	M	I

TABLE C.48   MESSAGE Supported Header Fields Used with All Other Responses

Header	Sending to UA	Receiving from UA	Use
Call-ID	M	M	
Call-Info	M	M	Call information
Content-Disposition	M	I	
Content-Encoding	M	I	
Content-Language	M	I	
Content-Length	M	M	
Content-Type	M	I	
Cseq	M	M	
Date	M	M	Insertion of date in requests and responses
From	M	M	
MIME-Version	M	I	
Organization	M	M	Organization identifier
P-Access-Network-Info	X	M	Access network identifier
P-Asserted-Identity	M	M	Asserted identity
P-Charging-Function-Addresses	M	M	Charging function addresses
P-Charging-Vector	M	M	Charging vector identifier
P-Preferred-Identity	X	M	Preferred identity
Privacy	M	M	Privacy mechanism
Require	M	M	Required methods identifier
Server	M	I	
Timestamp	I	I	
To	M	M	
User-Agent	M	I	
Via	M	M	
Warning	M	I	

NOTIFY Method

TABLE C.49   NOTIFY Supported Header Fields

Header	Sending to UA	Receiving from UA	Use
Accept	M	I	
Accept-Contact	M	I	Caller preferences
Accept-Encoding	M	I	
Accept-Language	M	I	
Allow	M	I	
Allow-Events	M	M	Event notification

TABLE C.49 NOTIFY Supported Header Fields (*continued*)

Header	Sending to UA	Receiving from UA	Use
Authorization	M	I	
Call-ID	M	M	
Contact	M	I	
Content-Disposition	M	I	
Content-Encoding	M	I	
Content-Language	M	I	
Content-Length	M	M	
Content-Type	M	I	
Cseq	M	M	
Date	M	M	Insertion of date in requests and responses
Event	M	M	
From	M	M	
Max-Forwards	M	M	
MIME-Version	M	I	
P-Access-Network-Info	X	M	Access network identifier
P-Asserted-Identity	M	M	Asserted identity
P-Charging-Function-Addresses	M	M	Charging function addresses
P-Charging-Vector	M	M	Charging vector identifiers
P-Preferred-Identity	X	M	Preferred identity
Privacy	M	M	Privacy mechanism
Proxy-Authorization	M	M	Authentication between UA and proxy
Proxy-Require	M	M	
Reason	M	I	Reason codes
Record-Route	M	M	Insertion of proxy address for routing
Reject-Contact	M	I	Caller preferences
Request-Disposition	M	I	Caller preferences
Require	M	M	Identifying required methods to be supported
Route	M	M	
Security-Client	X	M	Security mechanism
Security-Verify	X	M	Security mechanism
Subscription-State	M	I	
Supported	M	M	Identifying supported methods
Timestamp	M	I	
To	M	M	
User-Agent	M	I	
Via	M	M	

TABLE C.50   NOTIFY Supported Header Fields Used with 2*xx* Response

Header	Sending to UA	Receiving from UA	Use
Allow	M	I	
Authentication-Info	M	I	
Contact	M	I	
Record-Route	M	M	Insert proxy address for routing
Supported	M	I	

TABLE C.51   NOTIFY Supported Header Fields Used with 3*xx* or 485 Response

Header	Sending to UA	Receiving from UA	Use
Allow	M	I	
Contact	M	M	Deleting contact headers
Error-Info	M	I	
Supported	M	I	

TABLE C.52   NOTIFY Supported Header Fields Used with Response 401

Header	Sending to UA	Receiving from UA
Allow	M	I
Error-Info	M	I
Proxy-Authenticate	M	M
Supported	M	I
WWW-Authenticate	M	I

TABLE C.53   NOTIFY Supported Header Fields Used with Response 404, 413, 480, 500, 503, 600, or 603

Header	Sending to UA	Receiving from UA
Allow	M	I
Error-Info	M	I
Retry-After	M	I
Supported	M	I

TABLE C.54   NOTIFY Supported Header Fields Used with Response 405

Header	Sending to UA	Receiving from UA
Allow	M	I
Error-Info	M	I
Supported	M	I

TABLE C.55   NOTIFY Supported Header Fields Used with Response 407

Header	Sending to UA	Receiving from UA
Allow	M	I
Error-Info	M	I
Proxy-Authenticate	M	M
Supported	M	I
WWW-Authenticate	M	I

TABLE C.56   NOTIFY Supported Header Fields Used with Response 415

Header	Sending to UA	Receiving from UA
Accept	M	I
Accept-Encoding	M	I
Accept-Language	M	I
Allow	M	I
Error-Info	M	I
Supported	M	I

TABLE C.57   NOTIFY Supported Header Fields Used with Response 420

Header	Sending to UA	Receiving from UA	Use
Allow	M	I	
Error-Info	M	I	
Supported	M	I	
Unsupported	M	M	Identify methods not supported

TABLE C.58   NOTIFY Supported Header Fields Used with Response 421 or 494

Header	Sending to UA	Receiving from UA	Use
Allow	O	M	
Error-Info	O	O	
Security-Server	M	N/A	Security mechanism
Supported	M	M	

TABLE C.59   NOTIFY Supported Header Fields Used with Response 484

Header	Sending to UA	Receiving from UA
Allow	M	I
Error-Info	M	I
Supported	M	I

TABLE C.60　NOTIFY Supported Header Fields Used with Response 489

Header	Sending to UA	Receiving from UA	Use
Allow	M	I	
Allow-Events	M	M	Event notification
Error-Info	M	I	

TABLE C.61　NOTIFY Supported Header Fields Used with All Other Responses

Header	Sending to UA	Receiving from UA	Use
Call-ID	M	M	
Content-Disposition	M	I	
Content-Encoding	M	I	
Content-Language	M	I	
Content-Length	M	M	
Content-Type	M	I	
Cseq	M	M	
Date	M	M	Insertion of date in requests and responses
From	M	M	
MIME-Version	M	I	
P-Access-Network-Info	X	M	Access network identifier
P-Asserted-Identity	M	M	Asserted identity
P-Charging-Function-Addresses	M	M	Charging function addresses
P-Charging-Vector	M	M	Charging vector identifier
P-Preferred-Identity	X	M	Preferred identity
Privacy	M	M	Privacy mechanism
Require	M	M	Identify required methods to be supported
Server	M	I	
Timestamp	M	I	
To	M	M	
User-Agent	M	I	
Via	M	M	
Warning	M	I	

OPTIONS Method

TABLE C.62   OPTIONS Supported Header Fields

Header	Sending to UA	Receiving from UA	Use
Accept	M	I	
Accept-Contact	M	M	Caller preferences
Accept-Encoding	M	I	
Accept-Language	M	I	
Allow	M	I	
Allow-Events	M	M	Event notification
Authorization	M	I	
Call-ID	M	M	
Call-Info	M	M	
Contact	M	I	
Content-Disposition	M	I	
Content-Encoding	M	I	
Content-Language	M	I	
Content-Length	M	M	
Content-Type	M	I	
Cseq	M	M	
Date	M	M	Insertion of date in requests and responses
From	M	M	
Max-Forwards	M	M	
MIME-Version	M	I	
Organization	M	M	Organization identifier
P-Access-Network-Info	X	M	Access network identifier
P-Asserted-Identity	M	M	Asserted identity
P-Called-Party-ID	M	I	Called party identifier
P-Charging-Function-Addresses	M	M	Charging function addresses
P-Charging-Vector	M	M	Charging vector identifiers
P-Preferred-Identity	X	M	Preferred identity
P-Visited-Network-ID	M	M	Visitied network identifier
Privacy	M	M	Privacy mechanism
Proxy-Authorization	M	M	Authentication between UA and proxy

(Continued)

TABLE C.62 OPTIONS Supported Header Fields (*continued*)

Header	Sending to UA	Receiving from UA	Use
Proxy-Require	M	M	
Reason	M	I	Reason codes
Record-Route	M	O	Insertion of proxy address for routing
Reject-Contact	M	M	Caller preferences
Request-Disposition	M	M	Caller preferences
Require	M	M	Identify required methods to be supported
Route	M	M	
Security-Client	X	M	Security mechanism
Security-Verify	X	M	Security mechanism
Supported	M	M	Supported methods
Timestamp	M	I	
To	M	M	
User-Agent	M	I	
Via	M	M	

TABLE C.63 OPTIONS Supported Header Fields Used with Response 100

Header	Sending to UA	Receiving from UA	Use
Call-ID	M	M	
Content-Length	M	M	
Cseq	M	M	
Date	M	M	Insertion of date in requests and responses
From	M	M	
To	M	M	
Via	M	M	

TABLE C.64 OPTIONS Supported Header Fields Used with 2*xx* Response

Header	Sending to UA	Receiving from UA	Use
Accept	M	I	
Allow	M	I	
Authentication-Info	M	I	
Contact	M	I	
Record-Route	M	O	Insertion of proxy address for routing
Supported	M	I	

TABLE C.65   OPTIONS Supported Header Fields Used with Response 3xx or 485

Header	Sending to UA	Receiving from UA
Allow	M	I
Contact	M	M
Error-Info	M	I
Supported	M	I

TABLE C.66   OPTIONS Supported Header Fields Used with Response 401

Header	Sending to UA	Receiving from UA
Allow	M	I
Error-Info	M	I
Proxy-Authenticate	M	M
Supported	M	I
WWW-Authenticate	M	I

TABLE C.67   OPTIONS Supported Header Fields Used with Response 404, 413, 480, 486, 500, 503, 600, or 603

Header	Sending to UA	Receiving from UA
Allow	M	I
Error-Info	M	I
Retry-After	M	I
Supported	M	I

TABLE C.68   OPTIONS Supported Header Fields Used with Response 405

Header	Sending to UA	Receiving from UA
Allow	M	I
Error-Info	M	I
Supported	M	I

TABLE C.69   OPTIONS Supported Header Fields Used with Response 407

Header	Sending to UA	Receiving from UA
Allow	M	I
Error-Info	M	I
Proxy-Authenticate	M	M
Supported	M	I
WWW-Authenticate	M	I

TABLE C.70   OPTIONS Supported Header Fields Used with Response 415

Header	Sending to UA	Receiving from UA
Accept	M	I
Accept-Encoding	M	I
Accept-Language	M	I
Allow	M	I
Error-Info	M	I
Supported	M	I

TABLE C.71   OPTIONS Supported Header Fields Used with Response 420

Header	Sending to UA	Receiving from UA
Allow	M	I
Error-Info	M	I
Supported	M	I
Unsupported	M	M

TABLE C.72   OPTIONS Supported Header Fields Used with Response 421 or 494

Header	Sending to UA	Receiving from UA	Use
Allow	O	M	
Error-Info	O	O	
Security-Server	M	N/A	Security mechanism
Supported	M	M	

TABLE C.73   OPTIONS Supported Header Fields Used with Response 484

Header	Sending to UA	Receiving from UA
Allow	M	I
Error-Info	M	I
Supported	M	I

TABLE C.74   OPTIONS Supported Header Fields Used with All Other Responses

Header	Sending to UA	Receiving from UA	Use
Call-ID	M	M	
Call-Info	M	M	
Content-Disposition	M	I	
Content-Encoding	M	I	
Content-Language	M	I	
Content-Length	M	M	
Content-Type	M	I	
Cseq	M	M	
Date	M	M	Insertion of date in requests and responses
From	M	M	
MIME-Version	M	I	
Organization	M	M	
P-Access-Network-Info	X	M	Access network identifier
P-Asserted-Identity	M	M	Asserted identity
P-Charging-Function-Addresses	M	M	Charging function addresses
P-Charging-Vector	M	M	Charging vector identifier
P-Preferred-Identity	X	M	Preferred identity
Privacy	M	M	Privacy mechanism
Require	M	M	Required methods
Server	M	I	
Timestamp	M	I	
To	M	M	
User-Agent	M	I	
Via	M	M	
Warning	M	I	

REGISTER Method

TABLE C.75   REGISTER Supported Header Fields

Header	Sending to UA	Receiving from UA	Use
Accept	M	I	
Accept-Encoding	M	I	
Accept-Language	M	I	
Allow	M	I	

(Continued)

TABLE C.75   REGISTER Supported Header Fields (*continued*)

Header	Sending to UA	Receiving from UA	Use
Allow-Events	M	M	Event notification
Authorization	M	I	
Call-ID	M	M	
Call-Info	M	M	Call information
Contact	M	I	
Content-Disposition	M	I	
Content-Encoding	M	I	
Content-Language	M	I	
Content-Length	M	M	
Content-Type	M	I	
Cseq	M	M	
Date	M	M	
Expires	M	I	
From	M	M	
Max-Forwards	M	M	
MIME-Version	M	I	
Organization	M	M	Organization identifier
P-Access-Network-Info	X	M	Access network identifier
P-Charging-Function-Addresses	M	M	Charging function addresses
P-Charging-Vector	M	M	Charging vector identifier
P-Visited-Network-ID	M	M	Visited network identifier
Privacy	M	M	Privacy mechanism
Proxy-Authorization	M	M	Authentication between UA and proxy
Proxy-Require	M	M	
Reason	M	I	Reason codes
Request-Disposition	M	M	Caller preferences
Require	M	M	Required methods
Route	M	M	
Security-Client	X	M	Security mechanism
Security-Verify	X	M	Security mechanism
Supported	M	M	Supported methods
Timestamp	M	I	
To	M	M	
User-Agent	M	I	
Via	M	M	

TABLE C.76   REGISTER Supported Header Fields Used with Response 100

Header	Sending to UA	Receiving from UA
Call-ID	M	M
Content-Length	M	M
Cseq	M	M
Date	M	M
From	M	M
To	M	M
Via	M	M

TABLE C.77   REGISTER Supported Header Fields Used with 2xx Response

Header	Sending to UA	Receiving from UA	Use
Accept	M	I	
Accept-Encoding	M	I	
Accept-Language	M	I	
Allow	M	I	
Authentication-Info	M	I	
Contact	M	I	
P-Associated-URI	M	I	Associated identities
Path	M	I	Path extension
Supported	M	I	

TABLE C.78   REGISTER Supported Header Fields Used with 3xx or 485 Response

Header	Sending to UA	Receiving from UA	Use
Allow	M	I	
Contact	M	M	Contact address
Error-Info	M	I	
Supported	M	I	

TABLE C.79   REGISTER Supported Header Fields Used with Response 401

Header	Sending to UA	Receiving from UA
Allow	M	I
Error-Info	M	I
Proxy-Authenticate	M	M
Security-Server	X	N/A
Supported	M	I
WWW-Authenticate	M	I

TABLE C.80   REGISTER Supported Header Fields Used with Response 404, 413, 480, 486, 500, 503, 600, or 603

Header	Sending to UA	Receiving from UA
Allow	M	I
Error-Info	M	I
Retry-After	M	I
Supported	M	I

TABLE C.81   REGISTER Supported Header Fields Used with Response 405

Header	Sending to UA	Receiving from UA
Allow	M	I
Error-Info	M	I
Supported	M	I

TABLE C.82   REGISTER Supported Header Fields Used with Response 407

Header	Sending to UA	Receiving from UA
Allow	M	I
Error-Info	M	I
Proxy-Authenticate	M	M
Supported	M	I
WWW-Authenticate	M	I

TABLE C.83   REGISTER Supported Header Fields Used with Response 415

Header	Sending to UA	Receiving from UA
Accept	M	I
Accept-Encoding	M	I
Accept-Language	M	I
Allow	M	I
Error-Info	M	I
Supported	M	I

TABLE C.84   REGISTER Supported Header Fields Used with Response 420

Header	Sending to UA	Receiving from UA
Allow	M	I
Error-Info	M	I
Supported	M	I
Unsupported	M	M

TABLE C.85   REGISTER Supported Header Fields Used with Response 421 or 494

Header	Sending to UA	Receiving from UA	Use
Allow	O	M	
Error-Info	O	O	
Security-Server	M	N/A	Security mechanism
Supported	M	M	

TABLE C.86   REGISTER Supported Header Fields Used with Response 423

Header	Sending to UA	Receiving from UA
Allow	M	I
Error-Info	O	O
Min-Expires	M	I
Supported	M	I

TABLE C.87   REGISTER Supported Header Fields Used with Response 484

Header	Sending to UA	Receiving from UA
Allow	M	I
Error-Info	M	I
Supported	M	I

TABLE C.88   REGISTER Supported Header Fields Used with All Other Responses

Header	Sending to UA	Receiving from UA	Use
Call-ID	M	M	
Call-Info	M	M	Call information
Content-Disposition	M	I	
Content-Encoding	M	I	
Content-Language	M	I	
Content-Length	M	M	
Content-Type	M	I	
Cseq	M	M	
Date	M	M	
From	M	M	
MIME-Version	M	I	
Organization	M	M	Organization identifier
P-Access-Network-Info	X	M	Access network identifier
P-Charging-Function-Addresses	M	M	Charging function addresses

(Continued)

TABLE C.88   REGISTER Supported Header Fields Used with All Other Responses (*continued*)

Header	Sending to UA	Receiving from UA	Use
P-Charging-Vector	M	M	Charging vector identifiers
Privacy	M	M	Privacy mechanism
Require	M	M	Required methods
Server	M	I	
Timestamp	M	I	
To	M	M	
User-Agent	M	I	
Via	M	M	
Warning	M	I	

SUBSCRIBE Method

TABLE C.89   SUBSCRIBE Supported Header Fields

Header	Sending to UA	Receiving from UA	Use
Accept	M	I	
Accept-Contact	M	M	Caller preferences
Accept-Encoding	M	I	
Accept-Language	M	I	
Allow	M	I	
Allow-Events	M	M	Event notification
Authorization	M	I	
Call-ID	M	M	
Contact	M	I	
Content-Disposition	M	I	
Content-Encoding	M	I	
Content-Language	M	I	
Content-Length	M	M	
Content-Type	M	I	
Cseq	M	M	
Date	M	M	Insertion of date in requests and responses
Event	M	M	
Expires	M	I	
From	M	M	
Max-Forwards	M	M	
MIME-Version	M	I	

TABLE C.89 SUBSCRIBE Supported Header Fields (*continued*)

Header	Sending to UA	Receiving from UA	Use
Organization	M	M	Organization identifier
P-Access-Network-Info	X	M	Access network identifier
P-Asserted-Identity	M	M	Asserted identity
P-Called-Party-ID	M	I	Called party identifier
P-Charging-Function-Addresses	M	M	Charging function addresses
P-Charging-Vector	M	M	Charging vector identifiers
P-Preferred-Identity	X	M	Preferred identity
P-Visited-Network-ID	M	M	Visited network identifier
Privacy	M	M	Privacy mechanism
Proxy-Authorization	M	M	Authentication between UA and proxy
Proxy-Require	M	M	
Reason	M	I	Reason codes
Record-Route	M	M	Insertion of proxy address for routing
Reject-Contact	M	M	Caller preferences
Request-Disposition	M	M	Caller preferences
Require	M	M	Required methods
Route	M	M	
Security-Client	X	M	Security mechanism
Security-Verify	X	M	Security mechanism
Supported	M	M	Supported methods
Timestamp	M	I	
To	M	M	
User-Agent	M	I	
Via	M	M	

TABLE C.90 SUBSCRIBE Supported Header Fields Used with 2*xx* Response

Header	Sending to UA	Receiving from UA	Use
Allow	M	I	
Authentication-Info	M	I	
Contact	M	I	
Expires	M	I	
Record-Route	M	M	Insert of proxy address for routing
Supported	M	I	

TABLE C.91   SUBSCRIBE Supported Header Fields Used with Response 3*xx* or 485

Header	Sending to UA	Receiving from UA	Use
Allow	M	I	
Contact	M	M	Contact address
Error-Info	M	I	
Supported	M	I	

TABLE C.92   SUBSCRIBE Supported Header Fields Used with Response 401

Header	Sending to UA	Receiving from UA
Allow	M	I
Error-Info	M	I
Proxy-Authenticate	M	M
Supported	M	I
WWW-Authenticate	M	I

TABLE C.93   SUBSCRIBE Supported Header Fields Used with Response 404, 413, 480, 486, 500, 600, or 603

Header	Sending to UA	Receiving from UA
Allow	M	I
Error-Info	M	I
Retry-After	M	I
Supported	M	I

TABLE C.94   SUBSCRIBE Supported Header Fields Used with Response 405

Header	Sending to UA	Receiving from UA
Allow	M	I
Error-Info	M	I
Supported	M	I

TABLE C.95   SUBSCRIBE Supported Header Fields Used with Response 407

Header	Sending to UA	Receiving from UA
Allow	M	I
Error-Info	M	I
Proxy-Authenticate	M	M
Supported	M	I
WWW-Authenticate	M	I

TABLE C.96   SUBSCRIBE Supported Header Fields Used with Response 415

Header	Sending to UA	Receiving from UA
Accept	M	I
Accept-Encoding	M	I
Accept-Language	M	I
Allow	M	I
Error-Info	M	I
Supported	M	I

TABLE C.97   SUBSCRIBE Supported Header Fields Used with Response 420

Header	Sending to UA	Receiving from UA	Use
Allow	M	I	
Error-Info	M	I	
Supported	M	I	
Unsupported	M	M	Unsupported methods

TABLE C.98   SUBSCRIBE Supported Header Fields Used with Response 421 or 494

Header	Sending to UA	Receiving from UA	Use
Allow	O	M	
Error-Info	O	O	
Security-Server	M	N/A	Security mechanism
Supported	M	M	

TABLE C.99   SUBSCRIBE Supported Header Fields Used with Response 423

Header	Sending to UA	Receiving from UA
Allow	M	I
Error-Info	M	I
Min-Expires	M	I
Supported	M	I

TABLE C.100   SUBSCRIBE Supported Header Fields Used with Response 484

Header	Sending to UA	Receiving from UA
Allow	M	I
Error-Info	M	I
Supported	M	I

TABLE C.101   SUBSCRIBE Supported Header Fields Used with Response 503

Header	Sending to UA	Receiving from UA
Allow	M	I
Error-Info	M	I
Retry-After	M	I
Supported	M	I

TABLE C.102   SUBSCRIBE Supported Header Fields Used with All Other Responses

Header	Sending to UA	Receiving from UA	Use
Call-ID	M	M	
Content-Disposition	M	I	
Content-Encoding	M	I	
Content-Language	M	I	
Content-Length	M	M	
Content-Type	M	I	
Cseq	M	M	
Date	M	M	Insertion of date in requests and responses
From	M	M	
MIME-Version	M	I	
Organization	M	M	Organization identifier
P-Access-Network-Info	X	M	Access network identifier
P-Asserted-Identity	M	M	Asserted identifier
P-Charging-Function-Addresses	M	M	Charging function addresses
P-Charging-Vector	M	M	Charging vector identifier
P-Preferred-Identity	X	M	Preferred identity
Privacy	M	M	Privacy mechanism
Require	M	M	Required methods
Server	M	I	
Timestamp	M	I	
To	M	M	
User-Agent	M	I	
Via	M	M	
Warning	M	I	

UPDATE Method

TABLE C.103 UPDATE Supported Header Fields

Header	Sending to UA	Receiving from UA	Use
Accept	M	I	
Accept-Contact	M	I	Caller preferences
Accept-Encoding	M	I	
Accept-Language	M	I	
Allow	M	I	
Allow-Events	M	M	Event notification
Authorization	M	I	
Call-ID	M	M	
Call-Info	M	M	Call information
Contact	M	I	
Content-Disposition	M	M	Caller preferences
Content-Encoding	M	M	
Content-Language	M	M	
Content-Length	M	M	
Content-Type	M	M	
Cseq	M	M	
Date	M	M	Insertion of date in requests and responses
From	M	M	
Max-Forwards	M	M	
MIME-Version	M	I	
Organization	M	M	Organization identifier
P-Access-Network-Info	X	M	Access network identifier
P-Charging-Function-Addresses	M	M	Charging function addresses
P-Charging-Vector	M	M	Charging vector identifiers
Privacy	M	M	Privacy mechanism
Proxy-Authorization	M	M	Authentication between UA and proxy
Proxy-Require	M	M	
Reason	M	I	Reason codes
Record-Route	M	O	Insertion of proxy address for routing
Reject-Contact	M	I	Caller preferences

(Continued)

TABLE C.103   UPDATE Supported Header Fields (*continued*)

Header	Sending to UA	Receiving from UA	Use
Request-Disposition	M	I	Caller preferences
Require	M	M	Required methods identifier
Route	M	M	
Security-Client	X	M	Security mechanism
Security-Verify	X	M	Security mechanism
Supported	M	M	Supported methods
Timestamp	M	I	
To	M	M	
User-Agent	M	I	
Via	M	M	

TABLE C.104   UPDATE Supported Header Fields Used with 2*xx* Response

Header	Sending to UA	Receiving from UA	Use
Accept	M	I	
Accept-Encoding	M	I	
Accept-Language	M	I	
Allow	M	I	
Authentication-Info	M	I	
Contact	M	I	
Supported	M	I	

TABLE C.105   UPDATE Supported Header Fields Used with 3*xx* Response

Header	Sending to UA	Receiving from UA
Allow	M	I
Contact	M	M
Error-Info	M	I
Supported	M	I

TABLE C.106   UPDATE Supported Header Fields Used with Response 401

Header	Sending to UA	Receiving from UA
Allow	M	I
Error-Info	M	I
Proxy-Authenticate	M	M
Supported	M	I
WWW-Authenticate	M	I

TABLE C.107 UPDATE Supported Header Fields Used with Response 404, 413, 480, 486, 500, 503, 600, or 603

Header	Sending to UA	Receiving from UA
Allow	M	I
Error-Info	M	I
Retry-After	M	I
Supported	M	I

TABLE C.108 UPDATE Supported Header Fields Used with Response 405

Header	Sending to UA	Receiving from UA
Allow	M	I
Error-Info	M	I
Supported	M	I

TABLE C.109 UPDATE Supported Header Fields Used with Response 407

Header	Sending to UA	Receiving from UA
Allow	M	I
Error-Info	M	I
Proxy-Authenticate	M	M
Supported	M	I
WWW-Authenticate	M	I

TABLE C.110 UPDATE Supported Header Fields Used with Response 415

Header	Sending to UA	Receiving from UA
Accept	M	I
Accept-Encoding	M	I
Accept-Language	M	I
Allow	M	I
Error-Info	M	I
Supported	M	I

TABLE C.111 UPDATE Supported Header Fields Used with Response 420

Header	Sending to UA	Receiving from UA	Use
Allow	M	I	
Error-Info	M	I	
Supported	M	I	
Unsupported	M	M	Unsupported methods identifier

TABLE C.112 UPDATE Supported Header Fields Used with Response 421 or 494

Header	Sending to UA	Receiving from UA	Use
Allow	O	M	
Error-Info	O	O	
Security-Server	M	N/A	Security mechanism
Supported	M	M	

TABLE C.113 UPDATE Supported Header Fields Used with Response 485

Header	Sending to UA	Receiving from UA	Use
Allow	M	I	
Contact	M	C1	Contact address
Error-Info	M	I	
Supported	M	I	

TABLE C.114 UPDATE Supported Header Fields Used with All Other Responses

Header	Sending to UA	Receiving from UA	Use
Call-ID	M	M	
Call-Info	M	M	Call information
Content-Disposition	M	I	
Content-Encoding	M	I	
Content-Language	M	I	
Content-Length	M	M	
Content-Type	M	I	
Cseq	M	M	
Date	M	M	Insertion of date in requests and responses
From	M	M	
MIME-Version	M	I	
Organization	M	M	Organization identifier
P-Access-Network-Info	X	M	Access network identifier
P-Charging-Function-Addresses	M	M	Charging function addresses
P-Charging-Vector	M	M	Charging vector identifier
Privacy	M	M	Privacy mechanism
Require	M	M	Required methods identifier
Server	M	I	
Timestamp	M	I	
To	M	M	
User-Agent	M	I	
Via	M	M	
Warning	M	I	

Bibliography

3GPP TS 23.003, "Numbering, Addressing, and Identification," Release 1999

Aboba, B., M. Beadles, "The Network Access Identifier," RFC 2486, January 1999

Aboba, B., J. Wood, "Authentication, Authorization, and Accounting (AAA) Transport Profile," RFC 3539, June 2003

Arkko, J., V. Torvinen, G. Camarillo, A. Niemi, T. Haukka, "Security Mechanism Agreement for the Session Initiation Protocol (SIP)," RFC 3329, January 2003

Barnes, M., "An Extension to the Session Initiation Protocol (SIP) for Request History Information," RFC 4244, November 2005

Camarillo, G., "The Early Session Disposition Type for the Session Initiation Protocol (SIP)," RFC 3959, December 2004

Camarillo, G., G. Blanco, "The Session Initiation Protocol (SIP) P-User-Database Private Header (P-Header)," RFC 4457, April 2006

Camarillo, G., G. Blanco, "The Session Initiation Protocol (SIP) P-Profile-Key Private Header (P-Header)," RFC 5002, August 2007

Camarillo, G., H. Schulzrinne, "Early Media and Ringing Tone Generation in the Session Initiation Protocol (SIP)," RFC 3960, December 2004

Camarillo, G., W. Marshall, "Integration of Resource Management and Session Initiation Protocol (SIP)," RFC 3312, October 2002

Campbell, B., J. Rosenberg, H. Schulzrinne, C. Huitema, D. Gurle, "Session Initiation Protocol (SIP) Extension for Instant Messaging," RFC 3428, December 2002

Day, M., J. Rosenberg, H. Sugano, "A Model for Presence and Instant Messaging," RFC 2778, February 2000

Day, M., S. Aggarwal, G. Mohr, J. Vincent, "Instant Messaging/Presence Protocol Requirements," RFC 2779, February 2000

Donovan, S., "The SIP INFO Method," RFC 2976, October 2000

Ejzak, R., "Private Header (P-Header) Extension to the Session Initiation Protocol (SIP) for Authorization of Early Media," RFC 5009, September 2007

Elwell, J., "Connected Identity in the Session Initiation Protocol (SIP)," RFC 4916, June 2007

Garcia-Martin, M., E. Henrikson, D. Mills, "Private Header (P-Header) Extensions to the Session Initiation Protocol (SIP) for the 3rd Generation Partnership Project (3GPP)," RFC 3455, January 2003

Handley, M., V. Jacobson, "SDP: Session Description Protocol," RFC 2327, April 1998

Jennings, C., J. Peterson, M. Watson, "Private Extensions to the Session Initiation Protocol (SIP) for Asserted Identity Within Trusted Networks," RFC 3325, November 2002

Klyne, G., C. Newman, "Date and Time on the Internet: Timestamps," RFC 3339 July 2002

Levin, O., "Suppression of Session Initiation Protocol (SIP) REFER Method Implicit Subscription," RFC 4488, May 2006

Mahy, R., "A Message Summary and Message Waiting Indication Event Package for the Session Initiation Protocol (SIP)," RFC 3842, August 2004.

Mahy, R., B. Biggs, R. Dean, "The Session Initiation Protocol (SIP) REPLACES Header," RFC 3891, September 2004

Mahy, R., D. Petrie, "The Session Initiation Protocol (SIP) 'Join' Header," RFC 3911, October 2004

Marshall, W., "Private Session Initiation Protocol (SIP) Extensions for Media Authorization," RFC 3313, January 2003

Marshall, W., F. Andreasen, "Private Session Initiation Protocol (SIP) Proxy-to-Proxy Extensions for Supporting the PacketCable Distributed Call Signaling Architecture," RFC 3603, October 2003

Niemi, A., "Session Initiation Protocol (SIP) Extension for Event State Publication," RFC 3903, October 2004

Peterson, J., "A Privacy Mechanism for the Session Initiation Protocol (SIP)," RFC 3323, November 2002

Peterson, J., C. Jennings, "Enhancements for Authenticated Identity Management in the Session Initiation Protocol (SIP)," RFC 4474, August 2006

Petrack, S., L. Conroy, "The PINT Service Protocol: Extensions to SIP and SDP for IP Access to Telephone Call Services," RFC 2848, June 2000

Polk, J., "Extending the Session Initiation Protocol (SIP) Reason Header for Preemption Events," RFC 4411, February 2006

Roach, A.B., "Session Initiation Protocol (SIP) – Specific Event Notification," RFC 3265, June 2002

Roach, A.B., B. Campbell, J. Rosenberg, "A Session Initiation Protocol (SIP) Event Notification Extension for Resource Lists," RFC 4662, August 2006

Rosenberg, J., "Request Authorization Through Dialog Identification in the Session Initiation Protocol (SIP)," RFC 4538, June 2006

Rosenberg, J., "The Session Initiation Protocol (SIP) UPDATE Method," RFC 3311, September 2002

Rosenberg, J., H. Schulzrinne, "Reliability of Provisional Responses," RFC 3262, June 2002

Rosenberg, J., H. Schulzrinne, G. Camarillo, A. Johnston, J. Peterson, R. Sparks, M. Handley, E. Schooler, "SIP: Session Initiation Protocol," RFC 3261, June 2002

Rosenberg, J., H. Schulzrinne, P. Kyzivat, "Caller Preferences for the Session Initiation Protocol (SIP)," RFC 3841, August 2004

Rosenberg, J., H. Schulzrinne, P. Kyzivat, "Indicating User Agent Capabilities in the Session Initiation Protocol (SIP)." RFC 3840, August 2004

Schulzrinne, H., J. Polk, "Communications Resource Priority for the Session Initiation Protocol (SIP)," RFC 4412, February 2006

Sparks, R., "The Session Initiation Protocol (SIP) Refer Method," RFC 3515, April 2003

Sparks, R., "The Session Initiation Protocol (SIP) Referred-By Mechanism," RFC 3892, September 2004

Willis, D., B. Hoeneisen, "Session Initiation Protocol (SIP) Extension Header Field for Registering Non-Adjacent Contacts," RFC 3327, December 2002

Willis, D., B. Hoeneisen, "Session Initiation Protocol (SIP) Extension Header Field for Service Route Discovery During Registration," RFC 3608, October 2003

Index

1xx responses, 17, 27

2xx responses, 18, 27

3rd Generation Partnership Project (3GPP), 4, 117–119. *See also* IP Multimedia Subsystem

3xx responses, 19, 27

4xx responses, 27

5xx responses, 27

6xx responses, 27

100 TRYING response, 57

180 RINGING response, 57–58

181 CALL IS BEING FORWARDED response, 58

182 QUEUED response, 58

183 SESSION PROGRESS response, 58–59

200 OK response, 59, 105–106

202 ACCEPTED response, 59–60

300 MULTIPLE CHOICES response, 60

301 MOVED PERMANENTLY response, 60

302 MOVED TEMPORARILY response, 60–61

305 USE PROXY response, 61

380 ALTERNATIVE SERVICE response, 61

400 BAD REQUEST response, 61

401 UNAUTHORIZED response, 62

402 PAYMENT REQUIRED response, 62

403 FORBIDDEN response, 62

404 NOT FOUND response, 62–63

405 METHOD NOT ALLOWED response, 63

406 NOT ACCEPTABLE response, 63

407 PROXY AUTHENTICATION REQUIRED response, 63

408 REQUEST TIMEOUT response, 64

410 GONE response, 64

413 REQUEST ENTITY TOO LARGE response, 64

414 REQUEST URI TOO LONG response, 64

415 UNSUPPORTED MEDIA TYPE response, 64

416 UNSUPPORTED URI SCHEME response, 65

420 BAD EXTENSION response, 65

421 EXTENSION REQUIRED response, 65

423 INTERVAL TOO BRIEF response, 65

480 TEMPORARILY UNAVAILABLE response, 65–66

481 CALL/TRANSACTION DOES NOT EXIST response, 66

482 LOOP DETECTED response, 66

483 TOO MANY HOPS response, 67

484 ADDRESS INCOMPLETE response, 67

485 AMBIGUOUS response, 67

486 BUSY HERE response, 67

487 REQUEST TERMINATED response, 68

488 NOT ACCEPTABLE HERE response, 68

489 BAD EVENT response, 68

491 REQUEST PENDING response, 68

493 UNDECIPHERABLE response, 68–69

500 SERVER INTERNAL ERROR response, 69

501 NOT IMPLEMENTED response, 69

502 BAD GATEWAY response, 69

503 SERVICE UNAVAILABLE response, 69–70

504 SERVER TIMEOUT response, 70

505 VERSION NOT SUPPORTED response, 70

513 MESSAGE TOO LARGE response, 70

600 BUSY EVERYWHERE response, 71

603 DECLINE response, 71

604 DOES NOT EXIST ANYWHERE response, 71

606 NOT ACCEPTABLE response, 72

A

a headers, SDP session descriptions, 50

a=cat:<category> attribute, 52

a=charset:<character set> attribute, 53

a=framerate:<frame rate> attribute, 54

a=keywds:<keywords> attribute, 52

a=lang:<language> attribute, 53–54

a=orient:<whiteboard orientation> attribute, 53

a=ptime:<packet time> attribute, 52–53

a=quality:<quality> attribute, 54

a=rcvonly attribute, 53

a=rtpmap<payload type><encoding name>/<clock rate><encoding parameters> attribute, 52

a=sdplang:<SDP language> attribute, 53

a=sendrecv attribute, 53

a=tool:<name and version of tool> attribute, 52

a=type:<conference type> attribute, 53

abbreviated header fields, 28

ACCEPT ENCODING header, 30

ACCEPT header, 30, 63

ACCEPT LANGUAGE header, 30

access controls, 145–146

access lines, 88

access, secure. *See* security

ACCT-CALLED-URI parameter, 130

ACCT-CALLING-URI parameter, 130

ACCT-CHARGE-URI parameter, 129

ACCT-ROUTING-URI parameter, 130

ACK method
  client requests, 97–98
  defined, 26
  loose routing, 106–107
  strict routing, 110–111
  supported header fields for, 164, 208

active duration, 50

active IPS, 157

<address type>, SDP session description header, 48–49

addresses. *See also* Universal Resource Identifiers
  alternate, 18–19
  TO header, 39–40
  IMS billing, 123
  IP
    assignment of, 92–93
    DNS, 11–12
    URIs, 43–44

physical, 74–75
  public, 74–75

ALERT INFO header, 30

ALERT parameter, 32–33

ALLOW header, 30–31

alternate addresses, 18–19

amplification, 139–140

anomaly analysis, 155–156

ANSI SS7 cause codes, 84

answers, session, 24–25, 94–96

application media type, 51

application servers (AS), 11, 139, 149–150

application-based IDSs, 155

applications
  presence, 77
  push-to-talk, 120

AS (application servers), 11, 139, 149–150

attacks, network. *See* security

attributes, 50, 52–54

authentication. *See also* registrars
  client error class status codes, 62–63
  network security, 149–150
  private user identities, 44–46
  PROXY-AUTHENTICATE header, 36
  PROXY-AUTHORIZATION header, 36–37
  during registration process, 76–77
  within trusted domains, 142–144

AUTHENTICATION INFO header, 31

authorization. *See also* authentication; registrars
  in network security, 144
  P-Media-Authorization extension, 125

AUTHORIZATION header, 31

auto-answer capabilities, 120

availability, participant, 14

B

b headers, 49

bandwidth information headers, 49

bandwidth line, 88

Base64 values, 50

basic registration, 74–77

BCID (Billing correlation ID), 129

bearer traffic encryption, 148

BGCFs (breakout gateway control functions), 10, 81–82

Billing correlation ID (BCID), 129

billing, IMS, 122–123

binding process, 74–75

botnets, 140

bots, 140–141

BRANCH fields
  requests, 17
  server responses, 101
  VIA headers, 98

breaches, network. *See* security

breakout gateway control functions (BGCFs), 10, 81–82

busy line verification, 129

BYE method
  client requests, 97
  defined, 26
  loose routing, 107–108
  session termination, 112–113
  supported header fields for, 165–169, 209–213

C

c headers, 49

cable industry, 143

call control function, switches, 5, 7–8. *See also* media gateway control functions

call control layers, 152–156

call flow
  SIP, 56, 82, 84
  SS7 networks, 79–81

call forwarding, 58

CALL-ID header, 23–24, 31

CALL INFO header, 31

call session control functions (CSCFs), 89

call traces, 128

called party number, 87

calling party number, 87

calling party's category, 87

calls, returned, 34–35

CANCEL method
  481 Response, 66
  client requests, 97
  defined, 26
  session termination, 113
  stateful proxies, 17
  supported header fields for, 170–171, 213–215

capabilities, participant, 14–15

Carrier Identification Code (CIC), 88

case sensitivity, header fields, 28

category attributes, 52

cause codes, SS7, 82–86

CCF (charging collection function), 123

cell phones
  bots, 141
  P-Answer-State extension, 120
  USER AGENT header, 40

cellular telephones. *See* cell phones

certificates, 149

character set attributes, 53

charging collection function (CCF), 123

CIC (Carrier Identification Code), 88

circuit switching, point-to-point, 1–5

circular routing, 67

classes, status code, 27

classifications, header field, 28

clear text, 56, 135–136, 138

Clear values, 50

client error status codes, 27

client failure status codes, 61–69

client requests, 97–99

codes. *See* status codes

conference type attributes, 53

conferences, SDP, 46

confidentiality, 144

<connection address>, SDP session description header, 49

connection indicators, 87–88

connection information headers, 49

CONTACT headers
  300 Multiple Choices response, 60
  302 Moved Temporarily response, 60
  loose routing, 106
  overview, 32
  P-Associated-URI extension, 122
  REGISTER method, 75

CONTENT DISPOSITION header, 32–33

CONTENT LENGTH header, 33

CONTENT TYPE header, 33

CONTENT-TRANSFER-ENCODING header, 33

control media type, 51

cookies, 136–137

creator headers, 48

credentials, user agent, 31
CSCFs (call session control functions), 89
CSeq header, 33, 101
CT values, 49

D

data media type, 51
databases
 DNS, 11–12
 root, 12
 user, 126
 wireless networks, 6
 wireline networks, 4
DATE header, 33
date stamps, 137
DCS (distributed call signaling) architecture,
 128–131
decryption, 146–147
delayed packets, 2
denial of service (DoS) attacks, 139–141
descriptions, session, 97
dialogs
 IDs, 23–24, 98–99
 initiating, 93–96
 provisional status codes, 57
 stateful proxies, 16–17
digital switching, 1, 5
distributed call signaling (DCS) architecture,
 128–131
Domain Name Server (DNS)
 DoS attacks on, 139
 overview, 11–12
 poisoning, 138
 SIP routing, 102
domain names, 12, 43–44
domains
 404 Not Found response, 62–63
 trusted, 142–143
DoS (denial of service) attacks, 139–141
downloads, 144
duration, active, 50

E

e headers, 49
early media, 124
ECF (event charging function), 123
Electronic Numbering (ENUM), 12–13, 44, 82
e-mail address headers, 49
emergency interrupt, 129
emergency priority services, 36
encoding and syntax layer, 22
encryption
 493 Undecipherable responses, 68
 authentication and authorization, 149–150
 overview, 146–149
 prevention of message tampering, 138
 P-Visited-Network-ID extension, 128
encryption keys, 49–50
entities, SIP, 22
ENUM (Electronic Numbering), 12–13, 44, 82
Error Info header, 33
errors, syntax, 61
event charging function (ECF), 123
EVENT header, 34
event notification, 11, 77–79
event packages, 68
expiration times, 65
EXPIRE parameter, 32
EXPIRES header
 302 Moved Temporarily response, 61
 overview, 34
 REGISTER method, 75
extensions
 client error class status codes, 65
 documenting, 117
 overview, 116–117
 P-Access-Network-Info header, 119
 packet cable
  P-DCS-Billing-Info header, 129–130
  P-DCS-LAES header, 130–131
  P-DCS-OSPS header, 128–129
  P-DCS-Redirect header, 131
  P-DCS-Trace-Party-ID header, 128
 P-Answer-State header, 120
 P-Asserted-Identity header, 121, 125–126

P-Associated-URI header, 121–122
P-Called-Party-ID header, 122
P-Charging-Function-Address header, 122–123
P-Charging-Vector header, 123–124
P-Early Media header, 124
P-Media-Authorization header, 125
P-Profile-Key header, 125–126
proprietary, 17, 115–117
P-User-Database header, 126–127
P-Visited-Network-ID Header header, 127–128
SIP, 38
treating, 117–118

F

Financial entity ID (FEID), 129
forking proxies
    amplification attacks, 139–140
    client requests, 98
    overview, 18
    success class status codes, 59
formats, 51–52. *See also* messages
forward call indicators, 87
forwarding, call, 58
frame rate attributes, 54
fraud. *See* security
FROM header, 23–24, 34, 111

G

gateways, 3–4. *See also* media gateways (MGs);
        signaling gateways (SGs)
global failure status codes, 27, 70–72
GSM networks
    mobility, 6–7
    Universal Integrated Circuit Cards and SIM, 45

H

hackers. *See* security
handshakes, 24–25, 95–96
headers
    ACCEPT, 30
    ACCEPT ENCODING, 30
    ACCEPT LANGUAGE, 30

ACK method, 164, 208
ALERT INFO, 30
ALLOW, 30–31
AUTHENTICATION INFO, 31
AUTHORIZATION, 31
BYE method, 165–169, 209–213
CALL ID, 31
CALL INFO, 31
CANCEL method, 170–171, 213–215
case sensitivity, 28
CONTACT, 32
CONTENT DISPOSITION, 32–33
CONTENT LENGTH, 33
CONTENT TYPE, 33
CONTENT-TRANSFER-ENCODING, 33
CSeq, 33
DATE, 33
dialogs, 23–24
encryption, 146–149
Error Info, 33
EVENT, 34
EXPIRES, 34
extension, 116–117
FROM, 34
IN-REPLY-TO, 34–35
INVITE method, 172–177, 215–220
MAX-FORWARDS, 35
MESSAGE method, 178–181, 220–224
message waiting indication, 78–79
MIME VERSION, 35
MIN EXPIRES, 35
NOTIFY method, 182–186, 224–228
OPTIONS method, 186–191, 229–233
ORGANIZATION, 35
overview, 28–30
PRIORITY, 36
PRIVACY, 35
PROXY-AUTHENTICATE, 36
PROXY-AUTHORIZATION, 36–37
PROXY-REQUIRE, 37
RECORD-ROUTE, 37
REGISTER method, 191–196, 233–238
REPLY-TO, 37–38
REQUIRE, 38
RETRY-AFTER, 38
ROUTE, 38

headers (*Cont.*)
  SDP descriptors, 48–51
  SERVER, 38–39
  SUBJECT, 39
  SUBSCRIBE method, 196–201, 238–242
  SUPPORTED, 39
  TIMESTAMP, 39
  TO, 39–40
  UNSUPPORTED, 40
  UPDATE method, 201–204, 243–246
  user agent servers, 15–17
  USER-AGENT, 40
  VIA, 40–41
  WARNING, 41
  WWW AUTHENTICATE, 41–42
hijacking, 135–137
home location registers (HLRs), 7
home subscriber servers (HSSs), 126–127
hop counter, 87
host-based IDSs, 154–155
HSSs (home subscriber servers), 126–127
Hypertext Transport Protocol (HTTP), 22

I

i headers, 48–49
IAMs (initial address messages), 79–80, 86–89
IANA (Internet Assigned Numbers Authority), 117
ICID (IMS Charging Identity), 123
ICON parameter, 32
I-CSCFs (interrogating-call session control
    functions), 126–127
identities
  overview, 42–44
  private user identity, 44–45
  public user identity, 45–46
  service, 11
  session, 48
  subscriber, 121–122
IDs, 152–156
  dialog, 23–24, 94–95, 98–99
  token, 145
IETF (Internet Engineering Task Force), 117
IMEIs (International Mobile Equipment
    Identifiers), 40

impersonating servers, 137–138
IMS (IP Multimedia Subsystem). *See* IP
    Multimedia Subsystem
IMS Charging Identity (ICID), 123
IN (Intelligent Network) architecture, 4–5, 80–81
indexes, 22
indicators, 87–88
INFO method, 26
initial address messages (IAMs), 79–80, 86–89
IN-REPLY-TO header, 34–35
Integrated Digital Services Network
    (ISDN), 5
integrity, of data, 144
Intelligent Network (IN) architecture, 4–5, 80–81
interconnection, 3–4
International Mobile Equipment Identifiers
    (IMEIs), 40
Internet Assigned Numbers Authority
    (IANA), 117
Internet Engineering Task Force (IETF), 117
Internet Protocol (IP), 74, 79, 81, 89
interoperability, 115–116. *See also* extensions
Inter-Operator Identifiers (IOIs), 123–124
interrogating-call session control functions
    (I-CSCFs), 126–127
intrusion detection systems (IDSs), 152–156
intrusion protection systems (IPSs), 156–157
INVITE method
  Alert Info header, 30
  CALL ID header, 31
  client requests, 97
  defined, 25
  extensions, 118
  mapping IAM parameters to, 86–89
  P-DCS-OSPS extension, 129
  P-DCS-Trace-Party-ID extension, 128
  re-invites, 111–112
  supported header fields for, 172–177, 215–220
IOIs (Inter-Operator Identifiers), 123–124
IP (Internet Protocol), 74, 79, 81, 89
IP addresses
  assignment of, 92–93
  DNS, 11–12
  URIs, 43–44
IP Multicast method, 76

IP Multimedia Subsystem (IMS)
  accessing networks, 93
  extensions
    overview, 118–119
    P-Charging-Function-Addresses, 122–123
    P-Profile-Key, 126
    P-User-Database, 126–127
    P-Visited-Network-ID, 127–128
  overview, 4
IPsec, 147
IPSs (intrusion protection systems), 156–157
ISDN (Integrated Digital Services Network), 5

K

k headers, 49–50
keys, encryption, 49–50
keywords, 52

L

LAES-CONTENT parameter, 131
LAES-SIG parameter, 131
languages, 30, 53–54
latency, 2, 21
law enforcement, 130–131, 144
layering, security, 151–152
layers, SIP, 22
legacy signaling methods. *See* Signaling System #7
LENGTH header, 33
lists, route, 108, 150–151
location servers, 19
looping, 35
loose routing, 103–108

M

master databases, 12
MAX-FORWARDS header, 35
MAX-FORWARDS values, 17
media descriptions, 46–47, 51–52
media, early, 124
media gateway control functions (MGCFs)
    488 Not Acceptable Here response, 68
    overview, 10

and SS7 parameters, 86–89
    and STPs, 81–82
    VoIP network, 8
media gateway controllers (MGCs), 58, 93
media gateways (MGs)
    183 Session Progress responses, 58
    defined, 8
    overview, 9–10
    P-DCS-LAES extension, 131
media types
    in Intelligent Network architecture, 81
    SIP SDP, 88
MESSAGE method
    202 Accepted response, 59
    authentication and authorization, 150
    client requests, 97
    defined, 26
    encryption, 148
    supported header fields for, 178–181, 220–224
Message Transfer Part (MTP), 81
message waiting indication (MWI), 77–79
messages
    concept of dialogs, 23–25
    headers. *See* headers
    overview, 21–23
    requests, 25–26
    responses, 26–28
    Session Description Protocol, 47
methods. *See also individual methods by name*
    ACK Method, 164, 208
    Allow header, 30–31
    BYE Method, 165–169, 209–213
    CANCEL Method, 170–171, 213–215
    headers, 29–30
    INVITE Method, 172–177, 215–220
    MESSAGE Method, 178–181, 220–224
    NOTIFY Method, 182–186, 224–228
    OPTIONS Method, 186–191, 229–233
    REGISTER Method, 191–196, 233–238
    SDP session description k headers, 50
    server error class status codes, 69
    SIP, 25–26
    SUBSCRIBE Method, 196–201, 238–242
    supported, 15–16, 118
    UPDATE Method, 201–205, 243–246

MGCFs. *See* media gateway control functions
MGCs (media gateway controllers), 58, 93
MGs. *See* media gateways
migration to IP, 74, 79, 81, 89
MIME VERSION header, 35
MIN EXPIRES header, 35
mixed order headers, 28
mobile switching centers (MSCs), 6–7
mobility, 102–103
monitoring systems, 152–156
MSCs (mobile switching centers), 6–7
MTP (Message Transfer Part), 81
multimedia. *See* Session Description Protocol
MWI (message waiting indication), 77–79

N

NAIs (Network Access Identifiers), 44
name display, 4
names, in DNS, 11–12
nature of address indicators, 87
Network Access Identifiers (NAIs), 44
<network type>, SDP session description header,
    48–49
network-based IDSs, 154, 156
networks. *See also* packet networks; security
    accessing to establish sessions, 92–93
    architecture of
        SIP-specific entities, 13–19
        traditional voice network, 2–7
        voice over IP network, 7–13
    attacks on
        bots and DDoS attacks, 140–141
        denial of service and amplification, 139–140
        impersonating server, 137–138
        overview, 134–135
        registration hijacking, 135–136
        session hijacking, 136–137
        tearing down sessions, 138
    of bots, 140
    GSM, 6–7, 45
non-repudiation, 144
notification, event. *See* event notification
NOTIFY method
    defined, 26
    EVENT header, 34

message waiting indication, 78–79
    supported header fields for, 182–186, 224–228
number portability, 4–5
numbering plan indicators, 87
numbers, telephone. *See* telephone numbers

O

o headers, 48
offers, session, 24–25, 93–96
offsets, 50–51
operator ringback, 129
operator services position (OSPS), 128–129
OPTIONS method
    client requests, 97
    defined, 26
    extensions, 118
    supported header fields for, 186–191, 229–233
ORGANIZATION header, 35
OSPS (operator services position), 128–129
owner/creator headers, 48

P

p headers, 49
P-Access-Network-Info header, 119
packet cable extensions, 128–131
packet networks
    versus circuit switching, 7
    extensions, 118–119
    media gateways, 9
    overview, 1–2
P-Answer-State header, 120
parameters. *See* methods
participants, session, 14–15. *See also* sessions
P-Asserted-Identity header, 121, 125–126
passive probes, 154
P-Associated-URI header, 121–122
passwords
    and access controls, 145–146
    security, 134
    URIs, 43
P-Called-Party-ID header, 122
P-Charging-Function-Address header, 122–123
P-Charging-Vector header, 123–124

P-DCS-Billing-Info header, 129–130
P-DCS-LAES header, 130–131
P-DCS-OSPS header, 128–129
P-DCS-Redirect header, 131
P-DCS-Trace-Party-ID header, 128
P-Early Media header, 124
phone number headers, 49
phones, cell. *See* cell phones
physical addresses, 74–75
P-Media-Authorization header, 125
point-to-point circuit switching, 1–5, 79–80
poisoning, DNS, 138
port fields, 51
portability, number, 4–5
P-Profile-Key header, 125–126
presence applications, 77
PRIORITY header, 36
privacy, 144
PRIVACY header, 35
private user identity, 44–45
probing activities, 135, 154
profiles
    public user identities, 45–46
    traffic, 155–156
Prompt values, 50
property attributes, 50
proprietary extensions, 17, 115–117
protocol structure. *See also* headers
    identities, 42–46
    messages and formats
        concept of dialog, 23–25
        overview, 21–23
        requests, 25–26
        responses, 26–28
    SDP
        attributes, 52–54
        media descriptions, 51–52
        overview, 46–47
        session descriptions, 48–50
        time descriptions, 50–51
provisional status codes, 27, 57–59, 99
proxy servers. *See also* forking proxies
    client requests, 98
    emergency session establishment, 102
    headers, 42
    loose routing, 104–108

P-Access-Network-Info extension, 119
P-ASSERTED-IDENTITY extensions, 121
P-Called-Party-ID extension, 122
P-DCS-Billing-Info extension, 130
P-Media-Authorization extension, 125
P-Preferred-Identity extension, 125–126
P-Profile-Key extension, 126
provisional status codes, 57
P-User-Database extension, 126–127
P-Visited-Network-ID extension, 127–128
redirection status codes, 61
server error status codes, 69–70
SIP routing, 103
stateful proxies, 16–17
stateless proxies, 17–18
strict routing, 108–111, 150–151
VIA headers, 40–41
PROXY-AUTHENTICATE header, 36
PROXY-AUTHORIZATION header, 36–37
PROXY-REQUIRE header, 37
PSTN. *See* Public Switched Telephone Network
public addresses, 74–75
Public Switched Telephone Network (PSTN)
    called party number, 87
    calling party number, 87
    calling party's category, 87
    connection indicators, 87–88
    forward call indicators, 87
    hop counter, 87
    overview, 79–86
    P-DCS-OSPS extension, 128–129
    P-Early Media extension, 124
    transit network selection, 88
    user service information, 88–89
public user identity, 45–46
P-User-Database header, 126–127
push-to-talk applications, 120
P-Visited-Network-ID Header header, 127–128

Q

q parameter, 32
quality of service (QoS)
    packet networks, 2
    P-Media-Authorization extension, 125

queries, DNS, 12
queued calls, 58

R

r headers, 50–51
realms, 142–143
receive only attributes, 53
Record Keeping Server ID (RKSID), 129
RECORD-ROUTE headers, 37, 75, 108, 110–111
redirect servers, 18–19, 137–139
redirected responses, 122
redirection status codes, 27, 60–61
redirects. *See* P-DCS-Redirect header
REGISTER method
  401 Unauthorized response, 62
  client requests, 97
  defined, 25
  overview, 19
  P-Associated-URI extension, 122
  private user identities, 46
  registration process, 75–77
  renewing registration, 32
  supported header fields for, 191–196, 233–238
registrars
  401 Unauthorized response, 62
  amplification attacks, 139
  application servers, 11
  functions of, 7
  overview, 19
  P-Associated-URI extension, 121–122
  RECORD-ROUTE header, 37
  registration process, 73–76
  SIP routing, 103
registration
  authentication, 144
  basic, 74–77
  event notification, 77–79
  hijacking, 135–136
  interworking with PSTN
    called party number, 87
    calling party number, 87
    calling party's category, 87
    connection indicators, 87–88
    forward call indicators, 87
    hop counter, 87

    overview, 79–86
    transit network selection, 88
    user service information, 88–89
  overview, 73–74
  route lists, 108
  user, 14–15
re-invites, 111–112
release cause codes, 82–86
RENDER parameter, 32
repeat time headers, 50–51
REPLY-TO header, 37–38
REQUEST URI, 26, 67
requests. *See also* headers
  application servers, 11
  dialogs, 23–25
  first lines of, 55–56
  proxy servers, 16–18
  redirect servers, 18–19
  retransmission, 57
  user agent clients, 14–15
Requests for Comments (RFCs), 116–118, 159–162
request-URIs
  client requests, 98
  loose routing, 104–105
  P-Called-Party-ID extension, 122
  Privacy header, 35
  transit carriers, 88
REQUIRE header, 37–38
resolution, IP address, 12
Resource Reservation Protocol (RSVP), 125
resources, 43
responses. *See also* headers; *individual responses*
    by name; status codes
  BYE headers, 166–169, 210–213
  CANCEL headers, 170–171, 214–215
  CONTACT header, 32
  dialogs, 23–25
  INVITE headers, 173–177, 217–220
  MESSAGE headers, 179–181, 222–224
  NOTIFY headers, 183–186, 226–228
  OPTIONS headers, 188–191, 230–233
  proxy servers, 16–18
  redirected, 122
  REGISTER headers, 193–196, 235–238
  request/response correlation, 98
  to session offers, 94–95

SUBSCRIBE headers, 198–201, 239–242

UPDATE headers, 203–205, 244–246

user agent servers, 15

RETRY-AFTER header, 38

returned calls, 34–35

RFCs (Requests for Comments), 116–118, 159–162

ringback tones, 30, 57–58, 129

ringtones, 30

RKSID (Record Keeping Server ID), 129

roaming, 127–128

root databases, 12

ROUTE header, 38, 108, 111

route lists, 108, 150–151

routers, 7–8, 16. *See also* proxy servers

routing

  circuit-switched networks, 3

  circular, 67

  forking proxies, 18

  loose, 103–108

  overview, 102–103

  RECORD-ROUTE header, 37–38

  stateful proxies, 16–17

  strict

    versus loose, 103

    overview, 108–111

    RECORD-ROUTE header, 37

    ROUTE header, 38

    and security, 150–151

  user agent clients, 15

  VIA headers, 40–41

  VoIP, 10

RSVP (Resource Reservation Protocol), 125

S

s headers, 48

SCPs (service control points), 11, 80–81

S-CSCFs (serving-call session control functions), 126–127

SDP. *See* Session Description Protocol

security

  of access, 3

  authentication and authorization, 149–150

  encryption, 146–149

  network attacks

    bots and DDoS attacks, 140–141

    denial of service and amplification, 139–140

    impersonating server, 137–138

    overview, 134–135

    registration hijacking, 135–136

    session hijacking, 136–137

    tearing down sessions, 138

  overview, 133–134, 141–145

  password and access controls, 145–146

  PROXY-AUTHENTICATE header, 36

  PROXY-AUTHORIZATION header, 36–37

  SERVER header, 39

  solutions

    IDSs, 152–156

    IPSs, 156–157

    overview, 151–152

  strict routing, 150–151

SECURITY-CLIENT headers, 148–149

SECURITY-SERVER headers, 148–149

SECURITY-VERIFY headers, 148–149

send and receive mode attributes, 53

server error status codes, 27

server failure status codes, 69–70

SERVER header, 38–39

servers

  application, 139, 149–150

  impersonating, 137–138

  location, 19

  proxy

    forking proxies, 18

    stateful proxies, 16–17

    stateless proxies, 17–18

  redirect, 18–19, 137–139

  responses, 99–102

service control points (SCPs), 11, 80–81

service identifiers, 11

service providers, 44

service switching point (SSP) function, 5

serving-call session control functions (S-CSCFs), 126–127

Session Description Protocol (SDP)

  attributes, 52–54

  client requests, 97

  Content Disposition header, 32

  contents mapped to SS7 USIs, 88

  encryption, 147

  media descriptions, 51–52

Session Description Protocol (SDP) (*Cont.*)
  overview, 46–47
  session descriptions, 48–50
  session offers, 94
  time descriptions, 50–51
session descriptions
  client requests, 97
  SDP, 47
session information headers, 48–49
session name headers, 48
SESSION parameter, 32
sessions
  accessing network, 92–93
  client requests, 97–99
  dialogs, 23–25
  hijacking, 136–137
  initiating dialogs, 93–96
  modifying, 111–112
  overview, 14, 91–92
  P-Early Media extension, 124
  requests, 26
  requirements for, 97
  routing
    loose routing, 103–108
    overview, 102–103
    strict routing, 108–111
  server response, 99–102
  tearing down, 138
  terminating, 112–113
SGs (signaling gateways), 8, 10
signal transfer points (STPs), 80–81
signaling function, switches, 5
signaling gateways (SGs), 8, 10
Signaling System #7 (SS7)
  call flow, 79–81
  cause codes, 82–86
  establishing sessions with SIP domain, 86–89
  overview, 4
  SDP, 47
  signaling gateways, 10
Signaling Transport (SIGTRAN) protocol, 81
signature analysis, 155–156
SIGTRAN (Signaling Transport) protocol, 81
Simple Mail Transport Protocol (SMTP), 22
SIMs (Subscriber Identity Modules), 45
SIP Secure (SIPS), 147

SIP URIs, 42
SIP VERSION, request, 26
SIPS (SIP Secure), 147
SIPS URIs, 42
SMTP (Simple Mail Transport Protocol), 22
software versions
  SERVER header, 38–39
  USER AGENT header, 40
SS7. *See* Signaling System #7
SSP (service switching point) function, 5
start lines, request, 26
start times, session, 50
stateful proxies
  emergency session establishment, 102
  versus forking proxies, 18
  overview, 16–17
  SIP routing, 103
  strict routing, 150–151
stateless proxies
  overview, 17–18
  SIP routing, 103
status codes
  client failure, 61–69
  global failure, 70–72
  overview, 27, 55–56
  provisional, 57–59
  redirection, 60–61
  REGISTER headers, 196
  server failure, 69–70
  SS7 cause codes, 82–86
  successful, 59–60
status lines, response, 26–27
stop times, session, 50
STPs (signal transfer points), 80–81
strict routing
  versus loose, 103
  overview, 108–111
  RECORD-ROUTE header, 37
  ROUTE header, 38
  and security, 150–151
SUBJECT header, 36, 39
SUBSCRIBE method
  202 Accepted response, 60
  defined, 26
  event notification, 77
  supported header fields for, 196–201, 238–242

Subscriber Identity Modules (SIMs), 45
subscribers. *See also* participants, session
  changes in registration, 77
  global failure status codes, 71–72
  location of in wireless networks, 7
  message waiting indication, 77–79
  public user identity, 45–46
  redirection class status codes, 60–61
  registration hijacking, 135–136
  unavailable, 65–66
subscriptions
  nonexistent, 66
  private identities, 44–45
successful status codes
  200 OK response, 59
  202 ACCEPTED response, 59–60
  defined, 27
SUPPORTED header, 39
supported methods, 15–16, 30–31
switchboards, 1
switches
  network elements in VoIP, 7–8
  overview, 1–5
  sessions, 91–92
  SS7 networks, 79–81
switching fabric, 5. *See also* media gateways
syntax and encoding layer, 22
syntax errors, 61

T

t headers, 50
TAG parameters
  dialog IDs, 94–95
  FROM header, 34
  response, 23–24, 59
tearing down, sessions, 138
TEL URIs, 43–44, 82
telecommunications, 1–2
telephone numbers, 12–13, 43–44
telephones. *See* cell phones
text, clear, 56
text messages, 138
text values, 41
time descriptions, 50–51
timers, response, 98–99

TIMESTAMP header, 39
TLS (Transport Layer Security), 147
TO headers
  dialogs, 23–24
  overview, 39–40
  Privacy header, 35
  in routing, 111
  TAG field, 59
token IDs, 145
tool attributes, 52
traces, call, 128
traditional voice network
  wireless network architecture, 5–7
  wireline network architecture, 2–5
transactions, 22–23, 94
transit network selection, 88
transport fields, 51
transport layer, 22
Transport Layer Security (TLS), 147
trusted domains, 142–143
tunneling, 147–148

U

u headers, 49
UACs. *See* user agent clients
UAs. *See* user agents
UASs. *See* user agent servers
Universal Integrated Circuit Cards (UICCs), 45
Universal Resource Identifiers (URIs). *See also*
    request-URIs
  client error class status codes, 65
  CONTACT header, 32
  defined, 26
  overview, 42–44
  P-Associated-URI extension, 121–122
  request, 35, 88
  SDP session descriptions, 49
  in SIP routing, 102–103
  stateful proxies, 17
  TEL URIs, 43–44, 82
UNSUPPORTED header, 40
UPDATE method
  defined, 26
  session modification, 112
  supported header fields for, 201–205, 243–246

URIs. *See* Universal Resource Identifiers
user agent clients (UACs). *See also* client failure
    status codes
  client requests, 97–98
  dialog initiation, 93–95
  FROM header, 34
  loose routing, 104–108
  overview, 14–15
  P-Associated-URI extension, 121–122
  P-Media-Authorization extension, 125
  provisional status codes, 57–58
  redirection class status codes, 60
  registration process, 75
  re-invites, 111–112
  server response, 99–101
  session termination, 112–113
  strict routing, 108–111
  success class status codes, 59
user agent servers (UASs), 15–16
  dialog initiation, 93–95
  loose routing, 104–108
  P-Called-Party-ID extension, 122
  provisional status codes, 57–58
  re-invites, 111–112
  request/response correlation, 98
  server error class status codes, 69–70
  server response, 99–101
  session termination, 112–113
  strict routing, 108–111
user agents (UAs)
  Authorization header, 31
  dialogs, 23–25, 94
  methods and parameters, 163
  P-ASSERTED-IDENTITY extensions, 121
  P-DCS-Trace-Party-ID extension, 128
  UACs, 14–15
  UASs, 15–16
user databases, 126
user field, URIs, 43
user service information, 88–89
USER-AGENT header, 40
usernames, 48
users
  response to requests by, 100–101

session modification notification, 112
transaction, 23

V

v headers, 48
value attributes, 50
vendors, extension, 115–117
<version>, SDP session description header, 48
VIA headers
  client requests, 98
  loose routing, 104–106
  overview, 40–41
  stateful proxies, 16–17
visitor location register (VLR), 7
voice network, traditional
  wireless network architecture, 5–7
  wireline network architecture, 2–5
voice over IP (VoIP) network
  AS, 11
  DNS, 11–12
  ENUM, 12–13
  MG, 9–10
  MGCF, 10
  overview, 7–9
  signaling gateways, 10
voicemail platforms, 77–78
VoIP. *See* voice over IP

W

WARNING header, 41
Webinars, 46
whiteboard orientation attributes, 53
WiFi connections, 137
WiMax, 93
wireless networks, 5–7, 92–93
wireline networks, 1–5
WWW AUTHENTICATE header, 41–42

Z

z headers, 50